Behavior Genetic Approaches in Behavioral Medicine

PERSPECTIVES ON INDIVIDUAL DIFFERENCES

CECIL R. REYNOLDS, *Texas A&M University, College Station*
ROBERT T. BROWN, *University of North Carolina, Wilmington*

Current Volumes in This Series

BEHAVIOR GENETIC APPROACHES IN BEHAVIORAL MEDICINE
Edited by J. Rick Turner, Lon R. Cardon, and John K. Hewitt

COGNITIVE ASSESSMENT
A Multidisciplinary Perspective
Edited by Cecil R. Reynolds

DEAFNESS, DEPRIVATION, AND IQ
Jeffery P. Braden

DEMENTIA
Allen Jack Edwards

EXPLORATIONS IN TEMPERAMENT
Edited by Jan Strelau and Alois Angleitner

FIFTY YEARS OF PERSONALITY PSYCHOLOGY
Edited by Kenneth H. Craik, Robert Hogan, and Raymond N. Wolfe

HANDBOOK OF CREATIVITY
Assessment, Research, and Theory
Edited by John A. Glover, Royce R. Ronning, and Cecil R. Reynolds

INDIVIDUAL DIFFERENCES IN CARDIOVASCULAR RESPONSE TO STRESS
Edited by J. Rick Turner, Andrew Sherwood, and Kathleen C. Light

INTERNATIONAL HANDBOOK OF PERSONALITY AND INTELLIGENCE
Edited by Donald H. Saklofske and Moshe Zeidner

PERSONALITY, SOCIAL SKILLS, AND PSYCHOPATHOLOGY
An Individual Differences Approach
Edited by David G. Gilbert and James J. Connolly

SCHIZOPHRENIC DISORDERS
Sense and Nonsense in Conceptualization, Assessment, and Treatment
Leighton C. Whitaker

Behavior Genetic Approaches in Behavioral Medicine

Edited by

J. Rick Turner

Medical College of Georgia
Augusta, Georgia

Lon R. Cardon

Sequana Therapeutics
La Jolla, California

and

John K. Hewitt

Institute for Behavioral Genetics
University of Colorado
Boulder, Colorado

Springer Science+Business Media, LLC

Library of Congress Cataloging-in-Publication Data

Behavior genetic approaches in behavioral medicine / edited by J. Rick
Turner, Lon R. Cardon, and John K. Hewitt.
p. cm. -- (Perspectives on individual differences)
Includes bibliographical references and index.
ISBN 978-1-4757-9379-6 ISBN 978-1-4757-9377-2 (eBook)
DOI 10.1007/978-1-4757-9377-2
1. Psychophysiology--Genetic aspects. 2. Behavior genetics.
3. Health risk assessment. 4. Health behavior--Genetic aspects.
I. Turner, J. Rick. II. Cardon, Lon R. III. Hewitt, John K.
IV. Series.
[DNLM: 1. Genetics, Behavioral. QH 457 B4184 1995]
R726.5.B379 1995
616'.042--dc20
DNLM/DLC
for Library of Congress 95-19341
CIP

ISBN 978-1-4757-9379-6

Originally published by Plenum Press, New York in 1995
Softcover reprint of the hardcover 1st edition 1995
10 9 8 7 6 5 4 3 2 1

Contributors

Dorret I. Boomsma, Department of Psychophysiology, Free University of Amsterdam, 1081 HV Amsterdam, The Netherlands

Lon R. Cardon, Sequana Therapeutics, 11099 North Torrey Pines Road, Suite 160, La Jolla, California 92037

Andrew C. Heath, Department of Psychiatry, Washington University School of Medicine, St. Louis, Missouri 63110

John K. Hewitt, Institute for Behavioral Genetics, University of Colorado at Boulder, Boulder, Colorado 80309

Miriam R. Linver, School of Family and Consumer Resources, University of Arizona, Tucson, Arizona 85721

Pamela A. F. Madden, Department of Psychiatry, Washington University School of Medicine, St. Louis, Missouri 63110

Matt McGue, Department of Psychology, University of Minnesota, Minneapolis, Minnesota 55455

Joanne M. Meyer, Department of Human Genetics, Medical College of Virginia, Virginia Commonwealth University, Richmond, Virginia 23298

Toni P. Miles, Center for Special Populations and Health, College of Health and Human Development, Pennsylvania State University, University Park, Pennsylvania 16802

Robert Plomin, Centre for Social, Genetic, and Developmental Psychiatry, Institute of Psychiatry, London, United Kingdom SE5 8AF

Chandra A. Reynolds, Institute for Behavioral Genetics, University of Colorado, Boulder, Colorado 80309

David C. Rowe, School of Family and Consumer Resources, University of Arizona, Tucson, Arizona 85721

Harold Snieder, Department of Psychophysiology, Free University of Amsterdam, 1081 HV Amsterdam, The Netherlands

Jeanine R. Spelt, Department of Clinical Psychology, Virginia Commonwealth University, Richmond, Virginia 23284

J. Rick Turner, Departments of Pediatrics and Preventive Medicine, University of Tennessee, Memphis, Tennessee 38105; *present address:* Department of Pediatrics and Georgia Prevention Institute, Medical College of Georgia, Augusta, Georgia 30912

Lorenz J. P. van Doornen, Department of Psychophysiology, Free University of Amsterdam, 1081 HV Amsterdam, The Netherlands

Keith E. Whitfield, Department of Biobehavioral Health, College of Health and Human Development, Pennsylvania State University, University Park, Pennsylvania 16802

Preface

Behavioral medicine is a wide field of investigation, with an emphasis on interdisciplinary collaboration. Since being founded in 1979, the Society of Behavioral Medicine has accordingly attracted scientists from many disciplines, including psychology, medicine, nursing, epidemiology, and public health. Increasingly, researchers from the discipline of behavior genetics are also becoming involved in behavioral medicine research. This volume showcases the contributions that behavior genetics can make in the study of health and disease. While discussions center around alcoholism, cardiovascular disease, obesity, and smoking behavior, these avenues of investigation should be regarded only as exemplars of areas in which twin and family studies can be highly informative. It is certainly the case that the approaches presented in this volume can be applied in many other areas of behavioral medicine research.

The invitation to prepare this volume, extended by Series Editor Robert T. Brown, came at a time of considerable interest in behavior genetic approaches in behavioral medicine. Accordingly, chapter authors enthusiastically contributed reports of current investigations. Formal collaboration between the disciplines of behavioral medicine and behavior genetics can be traced back—to a certain degree—to 1990, when the Society of Behavioral Medicine (SBM) and the Behavior Genetics Association jointly sponsored a symposium entitled "Quantitative Genetics and Behavioral Medicine," held at SBM's 11th Annual Meeting. During this symposium (facilitated by Dr. Michael Goldstein, then chair of SBM's Scientific Liaison Committee), six behavior geneticists shared their work on health-related topics with SBM's attendees. The symposium was well suited to the meeting's theme of "Nature, Nurture, and Behavioral Medicine." While this volume is certainly not simply a report of the symposium, it is noteworthy that several speakers and attendees are among those who have now provided accounts of the latest work in their fields.

The magnitude of genetic influences on behavior, termed "heritability," is estimated as the proportion of total behavioral (phenotypic) variance caused by genetic differences between individuals. Heritabilities are typically expressed as percentages. It has been shown for many behaviors that genetic influences contribute to individual variation. Such evidence is very important in theories of biological predispositions toward certain disease states; the diathesis–stress model postulates that stress precipitates illness where there is an existing vulnerability. However, heritabilities for many behaviors do not exceed 50%. While a figure of 50% indicates that genetic influences are important, it equally indicates that 50% of phenotypic variance cannot be explained by genetic influences. Thus, while other approaches leave the issue unresolved, it is often a properly designed behavior genetic study that most clearly demonstrates the importance of the environment.

Accordingly, one fundamental message that will be repeated several times in this volume deserves mention at the outset. It is understood widely, and correctly, that the analytical strategies of behavior genetics permit the study of genetic influences on individual variation in behavior (where the term "behavior" is used in its widest context). It has been shown, for example, that there is genetic influence on the degree of cardiovascular response during psychological stress. However, it is much less widely recognized that behavior genetic techniques are extremely powerful strategies for exploring environmental influences on behavior. As Plomin states in Chapter 12, "Genetics research provides the best available evidence for the importance of nonheritable factors." The importance of this latter capability of behavior genetic approaches cannot be overstated. As Cardon observes in Chapter 13, "From an environmental perspective, the study of twins or their families or both provides an exceptionally well-controlled design, a paradoxical asset of twins given their usual association with genetic hypotheses." Once heritable variation contributing to individual variation has been controlled for, the researcher has less total variation to explain, and the salient features of the environment can thus be evaluated with more precision.

The importance of both genetic and environmental influences on behavior may be of particular relevance to the discipline of behavioral medicine, which is concerned with prevention, diagnosis, treatment, and rehabilitation (Schwartz & Weiss, 1978). First, the clear existence of biological predispositions (genetic influences) allows the possibility of identifying individuals who may be most at risk for a given deleterious outcome. Second, the influence of environmental factors on behavior implies that successful intervention is certainly possible; as Plomin emphasizes in Chapter 12, heritability does not imply immutability or untreatability.

The chapters in this volume, then, describe investigations that demonstrate both genetic and environmental influences on health-related behaviors. The book is divided into seven parts. Following Part I, an introduction to behavior genetic approaches, Part II focuses on addictive behaviors, with particular attention being paid to alcohol and nicotine use. Part II describes studies that explore genetic and environmental determination of cardiovascular stress responses and examines developmental genetic trends in cardiovascular activity. Body weight, adiposity, and eating disorders are the focus of Part IV. Again, genetic and environmental influences across the life span are discussed.

Parts V and VI take a different approach. Rather than focusing on specific phenotypes, the chapters in Part V show how behavior genetic approaches can be particularly informative in the study of gender and ethnicity. For example, it is possible to see whether the same genes may be expressed differently in men and women. Also, ethnic groups may display differences in the degree of genetic determination of certain phenotypes. These differences may provide evidence for the increased impact of environmental risk factors when a group is exposed to them with greater frequency. The chapters in Part VI emphasize and detail the contribution of behavior genetic approaches to the study of the environment and introduce the reader to molecular genetic analyses that attempt to identify specific genes that influence complex behaviors and disorders. Finally in Part VII, Chapter 14 provides an integrative discussion that highlights the implications of a behavior genetic perspective in behavioral medicine.

The goal of this volume is to encourage the continued interfacing of behavior genetics and behavioral medicine in the study of health-related behaviors. It is perhaps noteworthy that the volume appears in a year when the theme of the SBM's annual meeting is "Stress–Diathesis: Behavioral Medicine and the Biological Predisposition of Disease." The chapters in this volume address both aspects of this theme. Stress may well precipitate disease when there is an existing vulnerability, or diathesis, but the existence of this vulnerability is not in itself sufficient to led to disease development. Both genetic and environmental influences are important in determining phenotype expression and hence the etiology of disease.

Sincere thanks are expressed to all authors for their willingness to comply with specific editorial requests. In addition, all authors have taken considerable care to make the contents of their chapters readily accessible to scientists in behavioral medicine who, like the first editor are not geneticists by training but who are excited by the tremendous possibilities afforded by behavior genetic approaches in the study of both genetic and environmental influences. Thanks are also expressed to Series Editor Robert T. Brown and Eliot Werner at Plenum for advice throughout the project; David Ragland and Len Syme for the academic freedom to instigate it; Bruce Alpert for allocating appropriate departmental resources during the main editorial phase; Judy Adams for administrative and secretarial assistance; and Robin Cook at Plenum for production of the published volume.

J. Rick Turner

REFERENCES

Schwartz, G. E., & Weiss, S. M. (1978). Behavioral medicine revisited: An amended definition. *Journal of Behavioral Medicine, 1,* 249–251.

Contents

PART I **ORIENTATION**

PART II **ADDICTIVE BEHAVIORS**

Behavior Genetic Approaches in Behavioral Medicine

PART ONE

ORIENTATION

CHAPTER ONE

Behavior Genetic Approaches in Behavioral Medicine

An Introduction

JOHN K. HEWITT AND J. RICK TURNER

INTRODUCTION

Behavior genetics is the study of the genetic *and* environmental determination of individual differences in characteristics with a behavioral component. Traditional genetics has treated *environmental* variation largely as "noise." Traditional epidemiology and many of the social sciences have either ignored *genetic* variation or wished it would go away. In some cases, the emphasis of these traditional disciplines on a particular component, *either* genetic *or* environmental, can serve a useful scientific purpose. In other cases, it may be misleading and an obstacle to understanding.

Thus, for example, the discovery that a particular genetic variant is more frequent in alcoholics would not mean that we have found the primary cause of alcoholism, although we may be on the track of a factor that contributes to the heritable component of the risk for alcoholism. Equally, to assume that individual differences in behavior, and their associated physical or health outcomes, *must* have an environmental etiology would be both to blind ourselves to the biological

---

JOHN K. HEWITT • Institute for Behavioral Genetics, University of Colorado at Boulder, Boulder, Colorado 80309. J. RICK TURNER • Departments of Pediatrics and Preventive Medicine, University of Tennessee, Memphis, Tennessee 38105; *present address:* Department of Pediatrics and Georgia Prevention Institute, Medical College of Georgia, Augusta, Georgia 30912.

Behavior Genetic Approaches in Behavioral Medicine, edited by J. Rick Turner, Lon R. Cardon, and John K. Hewitt. Plenum Press, New York, 1995.

diversity that characterizes all species and to impede progress toward a fuller understanding of such pressing biobehavioral problems as addictions, obesity, and hypertension.

Clearly, a priori judgments about the genetic or environmental determination of individual differences have no place in science or clinical practice. The strength of behavior genetics is that it utilizes research methods that enable us to determine empirically the relative contributions of genetic variation to differences between individuals and, given this knowledge, to refine our understanding of the nature of the environmental influences that shape behavior, health, and disease. These research methods and their applications in behavioral medicine are the subject of this volume. This chapter provides an introduction to the kinds of approaches and terminology that will be encountered.

HERITABILITY AND GENETIC ANALYSIS

The characteristics of individuals that we observe are their *phenotypes*. There are some phenotypes for which we find individual differences only in unusual circumstances. For example, we each have one heart, one head, and so on. There are other phenotypes, including most if not all of those of interest to the behavioral and health sciences, for which individuals differ. Such phenotypic variation can be seen for alcohol use and alcoholism, cigarette smoking, adiposity and body mass, and cardiovascular responses to stress.

Within any population, there is genetic variation. With the exception of monozygotic (identical) twins, each of us is genetically unique. There is also environmental variation. The environments in which we develop differ from one another. An obvious question that then arises for any given characteristic is that of the extent to which the phenotypic variation observed in a population is a consequence of genetic variation within that population. This is a general question about the *heritability* of characteristics such as the body mass index, resting blood pressure, cardiovascular responses to stress, the motivation to use substances, or the vulnerability to develop substance tolerance, dependence, and associated problems. The complement to heritability, i.e., the extent to which variation is attributable to environmental factors, is sometimes referred to as *environmentality.*

Knowledge of whether a characteristic is highly heritable, moderately so, or hardly influenced by genes at all is important for three primary reasons. First, before we turn either to genetics or to family or individual environments in our search for risk factors, we want to know whether these factors are likely to be important. If the heritability of a characteristic is very low, we might want to invest greater effort in searching for the salient features of the environment than we do in trying to identify individual genes that determine the phenotype. Conversely, if the heritability is high, then it would be a mistake to ignore the possibility of genetic risk factors.

Second, as all the authors in this volume will agree, determining the heritability of a trait is an important step toward studying such issues as the extent and impact of the interactions and correlations between genetic and environmental influences (Wachs & Plomin, 1991); the causes of the multivariate associations among risks, vulnerabilities, protective factors, and outcomes (Kendler, Kessler, Neale, Heath, & Eaves, 1993); the genetic and environmental determinants of stability and change during development (Hahn, Hewitt, Henderson, & Benno, 1990) (see also Chapters 8 and 9); the extent of gender-specific expression of genetic and environmental influences (Cloninger, Christiansen, Reich, & Gottesman, 1978) (see also Chapter 10); and determining the genetic and environmental complexity of a composite phenotype such as smoking (see Chapters 3 and 4). These issues necessarily involve a knowledge of the heritability of individual traits at specific stages of development. Indeed, the set of questions of a more complex kind that it is sensible to ask will be largely determined by our initial broad understanding of the magnitude of genetic influences on the phenotypes in question.

Third, in practical terms, knowing whether individual differences in characteristics such as the body mass index or the liability to develop alcohol-associated problems are moderately or highly heritable may be of considerable help in understanding our propensities and vulnerabilities. For example, knowing, in general terms, that individual differences in the body mass index are in large part genetically determined may be helpful in a clinical or counseling setting, alleviating some of the guilt felt by individuals with particular body types. It also allows us to make better-informed judgments about the psychological investment needed to effect substantial changes in our body mass. A high heritability is not synonymous with immutability, but equally, traits with moderate to high heritabilities, such as cognitive abilities (McGue, Bouchard, Iacono, & Lykken, 1993), personality traits (Bouchard, 1994), or the body mass index (Stunkard, Foch, & Hrubec, 1986), seem to be relatively resistant to substantial and durable change through environmental influence unless there are strenuous intervention efforts involving a more or less permanent restructuring of the environment. Of course, the body mass index will change across the life span in the absence of deliberate intervention, and there is now evidence that these changes are themselves under genetic control (Fabsitz, Carmelli, & Hewitt, 1992).

CAVEATS ABOUT HERITABILITY

Since determining the heritability of a phenotype is often the first step in a behavior genetic analysis, considerable attention is paid to making this determination throughout the volume. However, there are some important caveats that we should establish at the outset. First, although there is a strong argument for paying attention to risk factors in proportion to their impact on phenotypic variation, this is not the whole story. The contribution of a factor to variation in the population is a function of two quantities: the impact of the factor in individual cases and the frequency with which it occurs in the population. Thus, rare genetic defects might contribute very little to variation in the population as a whole, even though they have a major impact on the individuals who inherit them. This is the case for a number of rare genetic abnormalities that cause mental retardation, for example. What is true for genes is equally the case for environments; the consequences of severely adverse family circumstances might be dramatic for members of the family, but might occur sufficiently rarely not to contribute much to the total variation of the population.

The corollary of this is that genetic or environmental risk factors that have a small impact in individual cases might make a large contribution at the population level if they occur for a major portion of the population. Thus, the proportions of variation accounted for are not an infallible guide to the impact that a given kind of risk factor might have in specific cases; in order to judge this impact, we need to know the prevalence of this risk factor in the population.

Second, heritabilities and other genetic and environmental quantities are functions of the impact of genetic variation and of environmental differences *for a given population in particular circumstances*. Heritabilities are *not* absolute properties of physical or behavioral characteristics. Consider three examples. The heritability of the body mass index is estimated to be higher in young adults than it is in children (see Chapters 7 and 8), and it is likely to be different again, and involve some different genes, in middle age (Fabsitz et al., 1992). The contribution of the shared family environment to initiating smoking is estimated to be greater in Australian men than in American men (Heath et al., 1993). Last, McGue (1993) (see also Chapter 2) discusses the possibility that the heritability of liability to alcoholism might differ in men and women. These examples are meant to serve notice that we fully *expect* that components of genetic and environmental variation might differ across cultures, genders, at different ages or stages of development, and at different times in our history. For example, risk factors for smoking may change once the health consequences have become known. Determining the nature of these changes provides further insights into the interplay of genetic and environmental influences. It is part of the business of behavior genetics to develop methods that are sensitive to these differences and that provide empirical tests to detect them.

Third, for most phenotypes of interest to behavioral medicine, it is only relatively recently that studies of sufficient power and with proper attention to adequate subject sampling have begun to be published. Even now, single studies involving laboratory measurements are unlikely to be large enough to yield precise statements about the causes of individual differences. Typically, such studies, taken one at a time, might appear to be contradictory, with one reporting a large genetic effect, the next a relatively small one. Clearly, if the sample sizes are relatively small (e.g., tens rather than hundreds or thousands of families), this contradiction might be exactly what we would expect from statistical considerations. Thus, we should not draw firm conclusions from any one study unless we are sure about the adequacy of its size, sample ascertainment, and assessment procedures. For this reason, in addition to providing detailed reports of individual studies, the chapters in this volume assemble and review evidence from a range of studies in order to arrive at robust conclusions about the roles of genes and environments.

TWIN, FAMILY, AND ADOPTION STUDIES

The starting point for our interest in genetic variation is probably the observation of familial resemblance, or aggregation, of physical or psychological characteristics. Pairs of siblings resemble each other more than do randomly chosen pairs of individuals, and children resemble their parents, on average, to a greater degree than they resemble randomly chosen adults. Such degrees of resemblance can be assessed in terms of correlation coefficients for quantitative measurements, such as blood pressure or weight, or in terms of concordance rates for discretely defined characteristics, such as a disease state or psychiatric diagnosis.

In clinical studies, the proband is the individual affected by a disease or condition that causes a family to be included in a study. The probandwise concordance rate is the probability that a relative of a given type will also be affected. *In the absence of family resemblance,* the concordance rate should be equal to the prevalence of the disease or condition. Thus, for commonly occurring conditions, such as having ever smoked cigarettes, this baseline concordance rate might be as high as 80% in some populations. Family resemblance would then be indicated by concordance rates that were even higher. On the other hand, for conditions that are relatively infrequent, such as a psychiatric disorder like schizophrenia with a 1% prevalence, a concordance rate of 10% might indicate substantial family resemblance.

Because such concordance rates must be interpreted in the context of the appropriate prevalence rates, they are sometimes used to estimate another type of correlation coefficient, called *tetrachoric* or *polychoric* correlations, on the assumption that a continuous, normally distributed, liability underlies the occurrence of the disease or condition. Such correlations take into account the frequency, or prevalence, of the condition in the population, and, where they are appropriate, they have the advantage of allowing more direct tests of the predictions of genetic and environmental hypotheses.

Whatever method we use, we typically find that there is positive family resemblance, and the resemblance is more strongly positive the closer the degree of family relationship. The problem is that in typical families, or at least ordinary nuclear families, we find that the degree of genetic relationship is confounded with the degree of environmental or social relationship. The mere observation of familial aggregation is therefore insufficient to allow the inference of a genetic or an environmental etiology. Instead, we need to study pairs of relatives in whom the degree of genetic relationship differs when the environmental resemblance is kept the same or, alternatively, the degree of environmental resemblance differs when the degree of genetic relationship is held constant.

The "natural experiment" of twinning affords a good approximation to the first kind of situation, and the adoption of children to be reared away or apart from their biological parents provides the second kind of situation. There are numerous other possibilities involving half siblings, stepfamilies, and so on that might be informative, but, by and large, both the statistical power and the conceptual clarity are greatest for research studies that start with a nucleus of either twins or families involved in adoptions. It is therefore these two kinds of studies that form the great majority of behavior genetic studies of relevance to behavioral medicine.

To illustrate the logic and application of these approaches, we can consider the classic twin study, which builds on the biological fact that there are, genetically, two kinds of twins that provide contrasting degrees of genetic relationship in siblings of the same age and family circumstances. Thus, monozygotic (MZ) twins, or identical twins, have identical copies of all of their genes, while dizygotic (DZ) twins, or fraternal twins, share, on average, only half their genes identically by descent, just like ordinary full siblings.

If there are genetic influences on the phenotype, the MZ correlation will exceed that for the DZ twins. The greater the influence of the genes in determining individual differences in the phenotype, i.e., the greater the proportion of phenotypic variance attributable to genetic differences (the heritability), the greater will be the difference between the MZ and DZ correlations. Indeed, with some simplifying assumptions, twice the difference between the MZ and DZ correlations can be taken as an estimate of the heritability of the trait (Falconer, 1989).

If genetic influences are the only cause of familial aggregation, if there is no nonadditive genetic variation, and if mating is random with respect to the characteristics under study, then the correlation for DZ twins should be half that for MZ twins. If there are significant nonadditive genetic influences, such as dominance or interactions of several genes at different genetic loci (called *epistasis*), or if there is significant competition between the twins or other contrast effects that accentuate the genetic differences between siblings, then the DZ correlation may be less than half the MZ correlation. Conversely, either assortative mating (like marrying like) or imitative or cooperative effects within the sibship, e.g., cooperative involvement in smoking or drinking, may cause the DZ correlation to exceed half the MZ correlation.

Environmental influences on the phenotype have two distinct characteristic consequences. First, if there are environmental influences that are shared by siblings growing up in the same home, such as might be the case for some consequences of socioeconomic status or parenting style, then the MZ correlation and the DZ correlation will reflect this source of familial aggregation to the same extent. In the absence of genetic influences, these shared environmental causes of familial aggregation would lead to equal MZ and DZ correlations; in the presence of genetic influences, the shared environment will raise the DZ correlation relative to the MZ correlation and, typically, the DZ correlation will be greater than half the MZ correlation. Second, if there are environmental influences that are unique to individuals, such as might be the case for personal life events, then familial aggregation will be attenuated and the MZ and DZ correlations will be reduced, although their relative magnitudes will continue to reflect the importance of genetic vs. shared environmental causes of familial aggregation.

Also, if familial influences on behavior (either genetic or shared environmental) are different for males and females, then the correlation for opposite-sex DZ pairs should be lower than those for same-sex pairs (Chapter Ten discusses gender effects in more detail). Similar kinds of reasoning can be applied to the comparison of the resemblance of pairs of adoptive siblings and pairs of biological siblings, of adoptees to their adoptive parents compared with adoptees to their biological parents, and so on.

The expectations for quantitative measures of family resemblance (such as variances, covariances, and correlations) can be specified in detail on the basis of quantitative genetic theory together with our hypotheses about how the environment might exert its influence. Formal statistical approaches may use a technique known as *path analysis,* a form of structural equation modeling, to express the impact on the phenotype of the genetic and environmental influences. Because these influences are not observed directly, they are sometimes called *latent variables*. Specific hypotheses about genetic and environmental variation may be summarized in terms of path diagrams. The hypothesis, or model, represented in a path diagram will be rejected if an appropriate goodness-of-fit test is significant (just as we reject a null hypothesis when our statistical test is significant). Alternatively, if the hypothesis represented in the path diagram is not rejected, this acceptance provides evidence that the set of genetic and environmental assumptions tested is one that adequately accounts for the observed data. We can then proceed to estimate heritabilities and other parameters of interest. (It should be kept in mind that this model-fitting procedure cannot prove that our hypothesis, or model, is correct [just as we cannot prove that a null hypothesis is correct]; it can only indicate that a given set of genetic and environmental assumptions adequately accounts for our observations.)

Because we do not expect wide familiarity with path analysis among our readers, we have tried to keep the use of such path diagrams to a minimum within this volume, although in fact several have crept in. For readers who are motivated to seek a thorough initiation into these techniques, we recommend the text by Neale and Cardon (1992) with eleven other contributing authors, all of whom are leading behavior geneticists, and five of whom are contributing authors to this volume. That text provides a full account of path analysis, model-fitting, maximum likelihood parameter estimation, and such techniques as developmental and multivariate genetic modeling, including the use of the Cholesky decomposition referred to in Chapter Eight of this volume.

In summary, behavior geneticists have taken advantage of the normal occurrence of twinning and adoptions to observe situations in which the confounding of genetic and environmental relationships, characteristic of ordinary families, is broken down. From these observations, analytical techniques have been developed that permit the detection and estimation of the important genetic and environmental contributions to phenotypic variation. The chapters in this volume draw heavily on what can be learned from these types of behavior genetic studies.

THE PHYSICAL BASIS OF GENETIC INHERITANCE

So far, we have discussed genetic and environmental influences largely in abstract statistical terms. However, the rise of genetics as a science during this century is attributable in large measure to the confirmation of the statistical properties of inheritance in the physical properties of the units of inheritance, the genes.

The genes are organized in pairs, having as their material basis the deoxyribonucleic acid (DNA) molecules carried in the pairs of chromosomes that make up the complement of the nucleus of each cell in the body. For example, the 46 chromosomes of the human karyotype represent 22 pairs of autosomal chromosomes along with the special case of the sex-determining pair. In humans, there are perhaps 100,000 genes located on the chromosomes. Any particular gene may be thought of as residing at a particular locus on a given chromosome. The gene found at a particular locus is not necessarily identical for different individuals or even for the two members of the pair of chromosomes; between one third and one half of the genetic loci show such genetic variation, or polymorphism, in man. The alternative forms of a gene at a particular locus are called *alleles*. If the two alleles residing on the pair of chromosomes are the same, the individual is said to be a *homozygote* at that locus. If they are different, the individual is a *heterozygote*.

When cells divide to produce more cells, by the process of mitosis during embryogenesis and later development, the chromosomes of the original cell are duplicated in the new cells. The exception to this situation is the genetic segregation that occurs during sexual reproduction. The gametes, i.e., sperms and eggs, are produced by the process of meiosis. Only one gene from each pair in the male parent cells ends up in the sperm, and only one from each pair of the female parent ends up in the egg. When the sperm fertilizes the egg, a new cell is formed that has received half its genes from each parent. With the exception of genes linked along the same chromosome, we may assume that determination of which member of the gene pair is passed on from each parent is effectively random and what happens at one locus has no effect on what happens at another.

Segregation analysis involves the examination of human family pedigrees for evidence that a phenotypic condition follows the pattern of occurrence that would be predicted from these genetic principles, together with knowledge of whether the condition occurs only in homozygotes for a particular allele (recessive gene expression) or in heterozygotes for this allele as well (dominant gene expression).

Linkage analysis exploits the physical linkage of genes along a chromosome to determine whether a particular phenotypic condition is inherited along with "marker" genes at loci that can be readily identified because of their physical or chemical properties. Because of recombination events between pairs of chromosomes, genetic linkage is not an all-or-none phenomenon. The closer on the chromosome the functional gene determining the phenotype is to the marker gene, the tighter will be the genetic linkage. Elaboration of these simple principles permits the localization of functional genes to specific chromosomes and specific sites on those chromosomes. If a marker locus is very close to a functional locus (i.e., if they are very tightly linked), then it may be that a particular allele at the marker locus tends to be associated with a particular allele (e.g., for a disorder) at the functional locus throughout a population. In this case, the loci are said to be in *linkage disequilibrium*. However, such disequilibrium is not necessary for successful linkage analysis, as all that is required is that particular alleles be associated within each pedigree, or family, under study.

THE SEARCH FOR INDIVIDUAL GENES

Given the availability of techniques such as genetic linkage analysis, the reader might be surprised to find that this volume is not a compendium of linkage studies. The first reason it is not is that many of the interesting questions about the genetic and environmental determination of individual phenotypes, and their developmental changes and multivariate associations, can be addressed through twin, family, and adoption studies, without any specific knowledge of the location of the genes involved (Hewitt, 1993). This has sometimes been presented as a criticism of behavior genetics by those who have said something to the effect that we should accept that genes influence human behavior only when behavior geneticists can point to the specific molecular basis for that influence. But apart from some reassurance that there is a physical basis for the genetic influences, it is not always clear what additional understanding comes simply from knowing that there is a gene on this or that chromosome near this or that marker. At a later stage, if the DNA sequence of the functional gene can be determined and its function understood, then this information would yield a real advance in our knowledge.

Second, there is no dispute that there are hundreds of major genetic abnormalities that have behavioral consequences (often mental retardation) as a pleiotropic effect of the principal metabolic or structural defect (McKusick, 1992). *Pleiotropy* is the common etiology, traceable to a single gene, of more than one phenotypic characteristic. However, most behavioral characteristics, like most physical characteristics, vary quantitatively rather than qualitatively throughout the population. Furthermore, such characteristics are likely to have complex etiologies involving several or many different genetic influences, together with several or many different environmental factors. Thus, the impact of individual genes that collectively contribute to heritabilities of up to 50% or more is likely to be relatively small and the detection of their individual contributions relatively difficult.

These difficulties, which have impeded the search for what are now known as quantitative trait loci (QTLs), are beginning to be overcome by technical advances in molecular genetics (Plomin, Owen & McGuffin, 1994). In Chapter Thirteen, Cardon discusses how these advances can be applied. The search for individual genes, or QTLs, will complement the gains in understanding that have come from the application of other behavior genetic approaches to issues in behavioral medicine.

CONCLUSIONS

Behavior genetics has developed methods of study that address issues about the genetic *and* environmental etiology of individual differences in characteristics that have a behavioral component. As such, they are particularly well suited to studies in behavioral medicine in which we are concerned with both biology and behavior.

Many of the questions that are addressed throughout this volume are of a heritability type, albeit that they often go beyond the simplest question of whether there is a genetic influence to questions about how the genetic influences operate on different components of the phenotypes, at different stages of development, and the extent to which these influences might be modified in different cultures or genders. It is important to note that the study of heritability also allows us to resolve more clearly the role of the environment.

Finally, we recognize that we will want to know which specific, individual genes contribute to heritable variation and how exactly they make their contribution. For the complex characteristics of interest to behavioral medicine, this level of understanding is not yet a reality. However, the technical obstacles to attaining this goal are now being overcome, and behavior genetic research will provide a solid empirical basis for the use of these advances in behavioral medicine.

REFERENCES

Bouchard, T. J. (1994). Genes, environment, and personality. *Science, 264*, 1700–1701.

Cloninger, C. R., Christiansen, K. O., Reich, T., & Gottesman, I. (1978). Implication of sex differences in prevalences of antisocial personality, alcoholism and criminality for familial transmission. *Archives of General Psychiatry, 35,* 941–951.

Fabsitz, R., Carmelli, D., & Hewitt, J. K. (1992). Evidence for independent genetic effects on obesity in middle age. *International Journal of Obesity, 16,* 657–666.

Falconer, D. S. (1989). *Introduction to quantitative genetics,* 3rd ed. Harlow, England: Longman.

Hahn, M. E., Hewitt, J. K., Henderson, N. D., & Benno, R. (1990). *Developmental behavioral genetics: Neural, biometrical, and evolutionary approaches.* New York: Oxford University Press.

Heath, A. C., Cates, R., Martin, N. G., Meyer, J., Hewitt, J. K., Neale, M. C., & Eaves, L. J. (1993). Genetic contribution to risk of smoking initiation: Comparisons across birth cohorts and across cultures. *Journal of Substance Abuse, 5,* 221–246.

Hewitt, J. K. (1993). The new quantitative genetic epidemiology of behavior. In R. Plomin & G. E. McClearn (Eds.), *Nature, nurture, and psychology* (pp. 401–415). Washington, DC: American Psychological Association.

Kendler, K. S., Kessler, R., Neale, M. C., Heath, A. C., & Eaves, L. J. (1993). The prediction of major depression in women: Toward an integrated etiologic model. *American Journal of Psychiatry, 49,* 109–116.

McGue, M. (1993). From proteins to cognitions: The behavioral genetics of alcoholism. In R. Plomin & G. E. McClearn (Eds.), *Nature, nurture, and psychology* (pp. 245–268). Washington, DC: American Psychological Association.

McGue, M., Bouchard, T. J., Iacono, W., & Lykken, D. T. (1993). Behavioral genetics of cognitive ability: A life-span perspective. In R. Plomin & G. E. McClearn (Eds.), *Nature, nurture, and psychology* (pp. 59–76). Washington, DC: American Psychological Association.

McKusick, V. A. (1992). *Mendelian inheritance in man: Catalogs of autosomal dominant, autosomal recessive, and X-linked phenotypes,* 10th ed. Baltimore: Johns Hopkins University Press.

Neale, M. C., & Cardon, L. R. (1992). *Methodology of genetic studies of twins and families.* Dordrecht, The Netherlands: Kluwer Academic Publishers.

Plomin, R., Owen, M., & McGuffin, P. (1994). The genetic basis of complex human behavior. *Science, 264,* 1733–1739.

Stunkard, A. J., Foch, T. T., & Hrubec, Z. (1986). A twin study of human obesity. *Journal of the American Medical Association, 256,* 51–54.

Wachs, T. D., & Plomin, R. (Eds.), (1991). *Conceptualization and measurement of organism environment interaction.* Washington, DC: American Psychological Association.

PART TWO

ADDICTIVE BEHAVIORS

CHAPTER TWO

Mediators and Moderators of Alcoholism Inheritance

Matt McGue

INTRODUCTION

One of the most consistent observations in the alcohol research field is that alcoholism recurs in families. Indeed, among behavioral disorders, alcoholism is remarkable for the strength of its familial loading. In her classic review of family studies of alcoholism, Cotton (1979) reported that alcoholics were approximately 6 times more likely than nonalcoholics to come from families with other alcoholic members. Although the extent to which the familial aggregation of alcoholism is influenced by genetic or environmental factors has long been a source of controversy in the alcohol research field (for reviews, see Murray, Clifford, & Gurling, 1983; Peele, 1986; Searles, 1988), and quantitative estimates are still imprecise and variable (Heath, Slutske, & Madden, 1994), empirical evidence now convincingly converges on the conclusion that genetic factors exert some influence on alcoholism etiology (McGue, 1994).

This chapter begins with a review of basic behavior genetic methodology and a discussion of how the application of this methodology serves to establish the existence of a genetic influence on alcoholism liability. Although there may be some who would still question the conclusion that genetic factors influence alcoholism risk, most scientists in the field now accept the existence of genetic influences and have thus redirected their research efforts toward trying to understand how, rather than simply whether, genes affect alcoholism (cf. Anastasi,

---

Matt McGue • Department of Psychology, University of Minnesota, Minneapolis, Minnesota 55455.

Behavior Genetic Approaches in Behavioral Medicine, edited by J. Rick Turner, Lon R. Cardon, and John K. Hewitt. Plenum Press, New York, 1995.

1958). Consequently, the second half of this chapter focuses on a review of research aimed at identifying and characterizing mechanisms of inherited influence.

BEHAVIOR GENETIC METHODOLOGY

SOURCES OF PHENOTYPIC VARIANCE

A major aim of behavior genetic analysis is the identification and characterization of the genetic and environmental factors that underlie individual differences in behavior. At least at the initial stages of inquiry, behavior geneticists have found it useful to consider three major contributors to between-individual variability: genetic factors, shared environmental factors (i.e., those environmental factors that are shared by reared-together relatives and thus contribute to their behavioral similarity), and nonshared environmental factors (i.e., those environmental factors that are not shared by reared-together relatives and thus contribute to their behavioral dissimilarity). Formally, the total phenotypic variance ($V\text{p}$, a measure of individual differences) is decomposed as

$$V\text{p} = V\text{g} + V\text{c} + V\text{e}$$

where $V\text{g}$, $V\text{c}$, and $V\text{e}$ represent, respectively, the genetic, shared environmental, and nonshared environmental components of phenotypic variance. It is often useful to represent the three major components of variance as percentages of the total phenotypic variance such that $h^2 = V\text{g}/V\text{p}$ gives the percentage of variance associated with genetic factors (i.e., the heritability) and $c^2 = V\text{c}/V\text{p}$ and $e^2 = V\text{e}/V\text{p}$ give, respectively, the percentages of variance associated with shared and nonshared environmental factors.

The utility of the additive decomposition of variance lies more with its ability to generate estimates that reasonably approximate the contribution of each of the components to total variance than with providing a wholly accurate portrayal of the true state of nature. It is unlikely that genetic and environmental factors act independently and additively as is assumed by this model. Nonetheless, approximate estimates of the strength of genetic and environmental contribution provide researchers with a useful starting point for designing studies aimed at characterizing the origins of individual differences in behavior.

Because reared-together relatives share both genes and an environment, phenotypic resemblance among them could be due to the effects of their shared genes, their shared environment, or both. Consequently, designs other than the study of intact nuclear families must be used to resolve the separate contribution of genetic and environmental factors to familial resemblance. Despite its simplicity, this observation seems to go unappreciated by many, including both biologically and psychosocially oriented researchers. Behavior geneticists have identified three general strategies for identifying the existence of genetic influences separate from the effects of a shared environment: adoption studies, twin studies, and genetic marker studies. Each of these methodologies has been used to assess a genetic contribution to alcoholism risk.

Adoption Study Design

The adoption study is elegant in the simplicity of its underlying logic, and it is for this reason that findings from adoption studies are often accorded the most weight in summary evaluations. Resemblance between a child reared from early infancy by a set of non-biologically related adoptive parents and that child's biological parents reflects the influence of genetic factors, under the assumption that the child shared no critical formative experience with his or her biological parents prior to placement in the adoptive home. Alternatively, resemblance between an adoptive child and his or her adoptive parents reflects the influence of shared environmental factors, under the assumption that placement in the adoptive home was independent of biological background.

Classic Twin Study Design

The inference of the existence of a genetic effect in a classic twin study is less direct than with an adoption study, although its natural extension of multivariate and developmental analyses makes the study of twins one of the most useful tools in human behavior genetics (Neale and Cardon, 1992). Identical twins, or monozygotic (MZ) twins, share 100% of their genetic material identical by descent, while nonidentical twins, or dizygotic (DZ) twins, share, on average, only 50% of their genetic material. Consequently, greater MZ than DZ twin similarity is indicative of genetic influence, but only if the two members of an MZ twin pair are no more likely than the two members of a DZ twin pair to share trait-relevant features of their environment (i.e., the equal environmental similarity assumption). Because the logic of the classic twin study depends so critically on the validity of this assumption, the equal environmental similarity assumption has received significant empirical attention in the behavior genetics literature (Loehlin & Nichols, 1976; Plomin, DeFries, & McClearn, 1990; Kendler, Neale, Keisler, Heath, & Evans, 1993).

Findings from this research can be summarized briefly. First, in apparent contradiction to the assumption, there is ample evidence that MZ twins are more likely to share experiences than are DZ twins (e.g., Vandenberg, 1976). In many cases, however, these differences in shared experience are unrelated to twin-pair behavioral similarity and so could not be the basis for greater MZ than DZ phenotypic similarity (Loehlin & Nichols, 1976). Moreover, in those cases in which there is an association between behavior and experience, experiential similarity may be more a consequence than a cause of behavioral similarity. For example, Lytton, Martin, and Eaves (1977) have shown that when mothers treat the two members of an MZ twin pair more similarity than they treat the two members of a DZ twin pair, they do so in large part because the MZ twins are more likely than the DZ twins to elicit similar reactions from their mothers. In the context of risk for alcoholism, Kendler, Heath, Neale, Kessler, and Eaves (1992) were unable to find any consistent relationship between environmental similarity, as children or as adults, and twin resemblance for alcoholism in adult female twins. Similarly, Heath, Jardine, and Martin (1989) found no such relationship for alcohol consumption.

Genetic Marker Studies

Adoption and twin studies attempt to identify the existence of genetic influences indirectly, by comparing risk rates among different classes of relatives. In contrast, genetic marker studies, by seeking to identify the specific genes that contribute to disease diathesis, provide a direct assessment of the existence of genetic influences. If genetic factors influence alcoholism risk, then one or more of the 50,000–100,000 genetic loci that comprise the human genome must be involved. Until relatively recently, attempts to identify the specific genes that contribute to risk for complex diseases were severely limited by the relatively small number of known genetic markers. However, the number of informative genetic markers has been increased substantially by the discovery of highly polymorphic noncoding segments of DNA (Botstein, White, Skolnick, & Davis, 1980).

Despite the increased sensitivity of genetic marker studies, however, these methods are still most likely to be successful with disorders in which the underlying genetic effect is due to one or a few, rather than many, contributing loci (Risch, 1990). Whether the genetic contribution to alcoholism risk is the result of just a few or of many genetic loci remains an open question. Consequently, the power of genetic marker studies with alcoholism is somewhat uncertain at this time.

There are two types of genetic marker studies, association studies and linkage studies. As association study aims to identify associations between genetic marker status and disease status at the *population* level. A classic example of an association is that between the rare joint disease ankylosing spondylitis and alleles within the human leukocyte antigen (HLA) system. Individuals who inherit the B27 HLA type are 88 times more likely to suffer ankylosing spondylitis than those who did not inherit this HLA type (Vogel & Motulsky, 1986). A positive association, if replicable, indicates that the genetic marker is either directly involved or located physically very near a gene that is directly involved in the etiology of the disorder.

A linkage study aims to identify association between genetic marker status and disease status *within families*. A positive linkage, if replicable, would indicate that a disease susceptibility locus exists within the same chromosomal region as the marker locus. Examples of linkage studies in human behavior genetics include reports of positive linkage of schizophrenia to chromosome 5 (Sherrington et al., 1988) and of bipolar affective disorder to chromosome 11

(Egeland, Gerhard, Pauls, Sussex, & Kidd, 1987). Unfortunately, positive linkage findings in human behavior genetics, including the two mentioned above, have proven to be difficult to replicate (McGuffin et al., 1990).

INHERITED BASIS OF ALCOHOLISM

Adoption Studies

There have been five different adoption studies of alcoholism, producing five comparisons for males and four comparisons for females. Table 1 summarizes results from these studies in terms of the rate of alcoholism among the reared-away offspring of alcoholic and non-alcoholic biological parents. Several general features of the table deserve comment. First, except for the early study by Roe (1944), studies of male adoptees are consistent in reporting a significantly

TABLE 1. Rates of Alcoholism among the Reared-Away Offspring of Alcoholics and Nonalcoholics

		History of alcoholism in biological parents		
Study	Country	Positive	Negative	Risk ratio[a]
Males				
Roe (1944)	United States	0.0% (N = 21)	0.0% (N = 11)	—
Goodwin, Schulsinger, Hermansen, Guze, & Winokur (1973)	Denmark	18.0% (N = 55)	5.0% (N = 78)	3.6[b]
Cloninger, Bohman, & Sigvardsson (1981)	Sweden	23.3% (N = 291)	14.7% (N = 571)	1.6[b]
Cadoret, O'Gorman, Troughton, & Heywood (1985)	United States	61.1% (N = 18)	23.9% (N = 109)	2.6[b]
Cadoret, Troughton, & O'Gorman (1987)	United States	62.5% (N = 8)	20.4% (N = 152)	3.1[b]
Females				
Roe (1944)	United States	0.0% (N = 11)	0.0% (N = 14)	—
Goodwin, Schulsinger, Knop, Mednick, & Guze (1977)	Denmark	2.0% (N = 49)	4.0% (N = 47)	0.5
Bohman, Sigvardsson, & Cloninger (1981)	Sweden	4.5% (N = 336)	2.8% (N = 577)	1.6[b]
Cadoret et al. (1985)	United States	33.3% (N = 12)	5.3% (N = 75)	6.3[b]

[a]The risk ratio is the rate of alcoholism among the reared-away offspring of alcoholics divided by the rate of alcoholism among the reared-away offspring of nonalcoholics. The risk ratio is not defined for the Roe study.
[b]The rate among reared-away offspring of alcoholics is significantly larger than the rate among reared-away offspring of nonalcoholics ($p < 0.05$).

higher rate of alcoholism among the reared-away offspring of alcoholics as compared to the reared-away offspring of nonalcoholics. The Roe study is notable not only in its failure to fit this general pattern, but also because Roe reported that none of the reared-away offspring of "problem drinkers" in her study experienced problems with alcohol in adulthood.

Although several researchers have legitimately questioned the methodological soundness of the Roe study (referring especially to the small size of the adoptee sample as well as to the ambiguous nature of the biological parent assessment and classification), the study is significant in suggesting that inherited factors may not be sufficient for the expression of alcoholism. That is, individuals with an inherited vulnerability to alcoholism need not express the disorder if reared in protective circumstances; for the Roe adoptees, such protection may have been afforded by being reared in rural settings during the Prohibition era.

A second notable feature of Table 1 is the large variance in reported rates of alcoholism across studies. For example, the rate of alcoholism among offspring of alcoholics varies more than 3-fold in males and more than 15-fold among females. Although this variance in risk likely reflects, in part, variance in alcoholism prevalence across time (Helzer, Canino, & Yeh, 1991) and place (Room, 1983), it also serves to highlight one of the major difficulties in undertaking research in this area, namely, assessment of the phenotype. Different diagnostic systems emphasize different aspects of problem drinking and therefore yield diagnoses with different prevalence rates. As a consequence, it is not easy to give a simple answer to the question: What is the rate of alcoholism among the offspring of alcoholics?

A final noteworthy feature of Table 1 is that the rate of alcoholism among the reared-away daughters of alcoholics is substantially lower than the rate of alcoholism among the reared-away sons of alcoholics. Moreover, with the possible exception of the single study by Cadoret, O'Gorman, Troughton, and Heywood (1985), the rate of alcoholism among the reared-away daughters of alcoholics differs minimally from the rate of alcoholism among the reared-away daughters of nonalcoholics. This pattern of findings suggests that genetic influences on alcoholism may be weaker for women than for men, a possibility considered later in the chapter.

Adoption studies also allow one to assess the environmental consequences of being reared by an alcoholic parent in the absence of the confounds of genetic inheritance. Despite likely screening for mental health, some adoptive parents are alcoholic. Consequently, a comparison of adoptees with and without an alcoholic adoptive parent provides a direct test of the effect of being reared by an alcoholic. Four of the adoption studies of alcoholism provide relevant comparisons, and these studies are summarized in Table 2. As can be seen, only one group, Cadoret and colleagues in Iowa, have reported increased rates of alcoholism among adoptees reared with alcoholics. Despite the exceptionable nature of Cadoret's finding, however, it is not easily dismissed, as it has been replicated by this research group in two separate studies and the effect has been observed with

TABLE 2. Rates of Alcoholism among Adoptees With and Without Adoptive Alcoholic Relatives

Study	Country	History of alcoholism in adoptive family: Positive	Negative	Risk ratio[a]
Males				
Goodwin et al. (1973)	Denmark	12.5% (N = 24)	10.1% (N = 109)	1.2
Cloninger et al. (1981)	Sweden	13.0% (N = 31)	18.0% (N = 831)	0.7
Cadoret et al. (1985)	United States	48.0% (N = 25)	24.5% (N = 102)	2.0[b]
Cadoret et al. (1987)	United States	38.5% (N = 26)	19.4% (N = 134)	2.0[b]
Females				
Bohman et al. (1981)	Sweden	3.7% (N = 27)	3.4% (N = 886)	1.1
Cadoret et al. (1985)	United States	17.4% (N = 23)	6.3% (N = 64)	2.8

[a]The risk ratio is the rate of alcoholism among the adoptees who have alcoholic adoptive relatives divided by the rate of alcoholism among adoptees who do not have adoptive alcoholic relatives. Except for the two studies by Cadoret, adoptive relatives refers to adoptive parents only.
[b]The rate among adoptees who have alcoholic adoptive relatives is significantly different from the rate among adoptees who do not have adoptive alcoholic relatives ($p < 0.05$).

both male and female samples (although in the latter case the 2-fold difference in rate of alcoholism does not quite attain statistical significance).

Two characteristics of Cadoret's methods and samples may help account for the apparent inconsistency of his findings with those by other adoption researchers. First, whereas the other studies summarized in Table 2 investigated the effect of having an alcoholic adoptive parent, Cadoret investigated the effect of being reared with any alcoholic adoptive relative. Perhaps the social modeling effects that others have posited to be critical in alcoholism etiology (Abrams and Niaura, 1987) are more potent among collateral (e.g., siblings) as compared to non-collateral relationships. Second, in at least one of the studies (Cadoret et al., 1985), the effect of having an adoptive alcoholic relative held among adoptees reared in rural, but not urban, settings, suggesting that familial modeling effects may be maximal in environments that provide the least opportunity for observing models of drinking outside the family.

Twin Studies

Table 3 summarizes findings from relevant twin studies of alcoholism in terms of the probandwise concordance rates [i.e., the risks to co-twins of affected probands (McGue, 1992)] for MZ and DZ twins. As with the adoption studies,

TABLE 3. Concordance for Alcoholism among Male and Female Same-Sex Twins

Study, country, and diagnosis	Twin concordance MZ	Twin concordance DZ	MZ/DZ ratio
Males			
Kaij (1960), Sweden			
Chronic alcoholism	0.71 (N = 14)	0.32 (N = 31)	2.2[a]
One or more temperance board registrations	0.61 (N = 59)	0.39 (N = 146)	1.6[a]
Hrubec & Omenn (1981), United States			
ICD-8 alcoholism	0.26 (N = 271)	0.12 (N = 444)	2.2[a]
Gurling, Oppenheim, & Murray (1984), England			
WHO alcohol dependence syndrome	0.33 (N = 15)	0.30 (N = 20)	1.1
Pickens et al. (1991),[b] United States			
DSM-III alcohol dependence	0.59 (N = 39)	0.36 (N = 47)	1.6[a]
DSM-III alcohol abuse and/or dependence	0.76 (N = 50)	0.61 (N = 64)	1.3[a]
McGue et al. (1992),[b] United States			
DSM-III alcohol abuse and/or dependence	0.77 (N = 85)	0.54 (N = 96)	1.4[a]
Caldwell & Gottesman (1991), United States			
DSM-III alcohol dependence	0.40 (N = 20)	0.13 (N = 15)	3.1[a]
DSM-III alcohol abuse and/or dependence	0.68 (N = 28)	0.46 (N = 26)	1.5[a]
Females			
Gurling et al. (1984), England			
WHO alcohol dependence syndrome	0.08 (N = 13)	0.13 (N = 8)	0.6
Pickens et al. (1991),[b] United States			
DSM-III alcohol dependence	0.25 (N = 24)	0.05 (N = 20)	5.0[a]
DSM-III alcohol abuse and/or dependence	0.36 (N = 31)	0.25 (N = 24)	1.4
McGue et al. (1992),[b] United States			
DSM-III alcohol abuse and/or dependence	0.39 (N = 44)	0.42 (N = 43)	0.9
Caldwell & Gottesman (1991), United States			
DSM-III alcohol dependence	0.29 (N = 7)	0.25 (N = 12)	1.2
DSM-III alcohol abuse and/or dependence	0.47 (N = 17)	0.42 (N = 24)	1.1
Kendler et al. (1992), United States			
DSM-III-R alcohol dependence	0.32 (N = 81)	0.24 (N = 79)	1.3[a]

[a] Twin concordance significantly different ($p < 0.05$).
[b] The Pickens et al. (1991) and McGue et al. (1992) samples overlap.

there are several notable features of the summary. First, for males, MZ twin concordance is consistently larger than DZ twin concordance, implicating the importance of genetic factors. Second, the MZ twin concordance for alcoholism is always substantially less than 1, implicating the importance of nonshared environmental factors. Finally, as was the case with adoption studies, findings from studies of female twins might be taken to suggest that genetic influences on alcoholism risk may be weaker among women than among men (McGue, 1993). That is, MZ twin concordance for alcoholism is substantially lower among females than among males. Moreover, several of the studies of female twins failed to find a significant difference in MZ and DZ concordance.

Genetic Marker Studies

There have been only a few linkage studies of alcoholism, and none of these studies has produced a strong or replicable linkage finding (Merikangas, 1990). In contrast, genetic variation in two systems has been implicated in alcoholism etiology by association studies. The first and most consistently associated system involves genetic polymorphisms for isoenzymes involved in the hepatic metabolism of alcohol [alcohol dehydrogenase (ADH)] and acetaldehyde [aldehyde dehydrogenase (ALDH)]. The second and most controversial association involves variation at the dopamine D_2 receptor (DRD_2) locus.

The principal metabolic pathway for ethanol elimination is oxidation by the liver enzyme ADH to acetaldehyde, which in turn is catalyzed by ALDH to acetate. Inherited variation in ADH and ALDH activity may help account for ethnic variation in alcoholism rates, particularly the relatively low rates seen among certain Asiatic populations. Variation in the activity of ALDH, the rate-limiting enzyme, is thought to be especially relevant to drinking behavior because elevated acetaldehyde levels are associated with acute toxic reactions to alcohol (e.g., the aversive therapeutic effect of disulfiram is achieved by inhibiting ALDH). Goedde, Harada, and Agarwal (1979) reported deficient ALDH activity in 50% of the livers of Chinese and Japanese donors. This reduction in activity was attributable to an inherited variant of ALDH2, the mitochondrial form of ALDH, that results from a single amino acid substitution (designated ALDH2-2). Both homozygotes and heterozygotes for ALDH2-2 show no detectable ALDH2 activity (Crabb, Edenberg, & Borson, 1987). The rate of deficient ALDH2 activity ranges from 30% to 50% among Japanese and Chinese, but is near zero among Europeans, Africans, and Native North Americans (Agarwal and Goede, 1989). Significantly, ALDH2 deficiency is clearly protective against alcoholism, as only 2.3% of Japanese alcoholics but nearly 50% of Japanese nonalcoholics are ALDH-deficient (Harada, Agarwal, Goedde, Tayaki, & Ishikawa, 1982).

Individuals with ALDH2 deficiency inherit a natural analog of the drug disulfiram so that their drinking is curtailed by the noxious "flushing response" they experience when they drink even a small amount of alcohol. Explaining the relationship between alcoholism and ALDH2 deficiency is complicated, however, by the observation of relatively high rates of alcoholism in populations with relatively high levels of ALDH2 deficiency. Specifically, the frequency of the ALDH2-2 allele is similar among Japanese, Chinese, and Koreans (Goedde, Agarwal, & Fritze, 1992). However, in contrast to the Japanese and Chinese, among whom population alcoholism rates are low, the rate of alcoholism among Koreans (at least male Koreans) is quite high. Indeed, in a recent multinational study that included samples from Korea, Taiwan, Canada, Puerto Rico, and the United States, the rate of alcoholism (as based on DSM-III criteria for Alcohol Abuse or Alcohol Dependence) was higher among Korean men than among any other demographic group (Helzer et al., 1990). Since ALDH2 status was not directly determined in this study, one cannot completely reject the possibility that a high rate of alcoholism characterizes only those Korean men who have not inherited ALDH2 deficiency. Nonetheless, these findings suggest again that inherited influences on alcoholism can be overcome by environmental factors (in this case the high-density drinking that appears to characterize certain cultural practices in Korea).

Experimental studies with animals have implicated the dopaminergic system in alcohol self-administration (e.g., Wise & Rompre, 1989) and have thus motivated a search for associations between polymorphisms that affect this system and rates of alcoholism in humans. The relevance of the dopaminergic system to human alcoholism seemed to be confirmed when Blum et al. (1990) reported that 69% of alcoholics but only 20% of nonalcoholics carried at least one copy of the A1 allele at a Taq1 restriction fragment length polymorphism site on the 3′ flanking region of DRD2. This striking association between the A1 allele and alcoholism reported by Blum et al. (1990) has since been replicated in two studies (Comings, et a., 1991; Parsian et al., 1991), but failed to be replicated in six others (Bolos et al., 1990; Gelernter et al., 1991; Schwab et al., 1991; Cook, Wang, Crowe, Hauser, & Freimer, 1992; Goldman et al., 1992; Turner et al., 1992). Moreover, one of the positive association studies (Parsian et al., 1991) failed to find linkage between markers of DRD2 and alcoholism.

One explanation for this inconsistent pattern of results derives from the observation that the positive association studies have generally been based on samples of severely affected alcoholics, while the negative association studies have generally been based on more broadly representative samples of alcoholics, leading to the hypothesis that the A1 allele may be a marker for risk of severe alcoholism only (Conneally, 1991). Nonetheless, the nature and significance of the association between the A1 allele and alcoholism are currently unclear. Attempts to specify a mechanism to explain the association are complicated by the fact that since the Taq1 site is downstream of the structural and regulatory segments of the DRD2 gene, the A1 allele has no clearly recognizable functional properties (Gelernter, Goldman, & Risch, 1993).

If there is an association between the A1 allele and alcoholism, then, it likely reflects one of two distinct possibilities. First, the A1 allele may exist in linkage disequilibrium with functional polymorphisms at DRD2, a possibility that is consistent with the population genetic data (Uhl, Blum, Noble, & Smith, 1993). Alternatively, the association may be an artifact of imperfect matching of the alcoholic and control samples. The frequency of the A1 allele is known to vary markedly across ethnic groups (Uhl et al., 1993), so that failure to carefully match the alcoholic and control groups for ethnic background could produce an artifactual association, a possibility long recognized with association studies (Vogel & Motulsky, 1986). Whatever the case, it is clear that the controversy surrounding the putative association between A1 and alcoholism portends the types of difficulties that researchers in the alcohol field will face as they attempt to move from twin and adoption studies, and our inferences from them about the existence and strength of genetic influences, to molecular genetic studies and attempts at more precise specifications of underlying genetic mechanisms.

Summary of Genetic Evidence

Findings from twin and adoption studies converge on the conclusion that genetic factors influence risk of alcoholism. These studies also underscore the importance of environmental factors: Concordance among MZ twins is far from perfect, and adoption studies hint at the effect of being reared in a family with a problem-drinking model. Given a genetic influence on alcoholism, genetic marker studies should be able to identify the specific genes involved. The search for single-gene effects on alcoholism is complicated, however, by the disorder's complex pattern of inheritance and clinical heterogeneity. Specifically, unlike the classic human genetic disorders for which molecular genetic methods have led to breakthrough discoveries, the inherited diathesis for alcoholism is likely to be polygenic and modifiable by cultural effects, two factors that are likely to greatly complicate molecular genetic research on alcoholism.

It is not surprising, then, that the specific genes that underlie risk for alcoholism, like the genes that underlie expression of most of the complex disorders, remain undiscovered. Even though their specific nature is unknown, however, all available clinical genetic data suggest that the genes that underlie alcoholism liability, when inherited, confer a vulnerability to, rather than the certainty of, alcoholism expression. That is, the inherited factors are highly sensitive to environmental modulation.

MECHANISMS OF INHERITED INFLUENCE

The conclusion that twin and adoption studies converge to establish the existence of a genetic influence on alcoholism, if accepted, motivates a shift in research from a past focus on trying to determine whether genes influence alcoholism to a present focus on trying to understand how genes might exert that influence. Molecular genetic studies aimed at identifying the specific genes that contribute to alcoholism risk represent one significant effort directed at characterizing the nature of genetic influence. But, as argued above, even though molecular genetic methods are ultimately likely to produce significant breakthroughs in our understanding of alcoholism etiology, the short-term success of these methods is likely to be limited by the complexity of the disorder. Consequently, molecular genetic approaches to alcoholism need to be complemented by studies aimed at identifying the physiological and behavioral manifestations of the genetic diathesis for alcoholism and determining how these factors combine or interact with environmental factors to increase or decrease risk of alcoholism. That is, there is a need for a top-down (behavior-to-gene) approach that complements the bottom-up (gene-to-behavior) approach of molecular genetic investigations.

In organizing alternative top-down efforts in alcoholism research, it is helpful to make use of the social psychological notion of mediating and moderating factors (Baron & Kenny, 1986). A mediating variable is defined as a variable that, because it is correlated with both, helps account for some or all of the statistical relationship between the independent and dependent variable. By extension, an inherited mediator would be a factor that is an intermediate phenotype in the pathway from primary gene product to overt disorder and thus helps account for the relationship between inherited factors and disease risk. Identifying inherited mediators can implicate specific causal mechanisms of inherited effects and can help refine molecular genetic efforts by identifying phenotypes that are more proximal to specific gene effects than alcoholism per se.

A moderator is a variable that influences the direction or strength of the relationship between the independent and dependent variable. By extension, a moderator of inheritance is a factor that either enhances or diminishes the influence of genetic factors on the behavioral phenotype. Identification of moderators of inheritance can increase the sensitivity of molecular genetic studies by helping to resolve etiological heterogeneity and can also provide a foundation for exploring the joint influence of genetic and environmental factors.

Mediators of Alcoholism Inheritance

An inherited mediator should be (1) correlated with risk for the disorder, (2) a cause rather than a consequence of the disorder, (3) itself inherited, and (4) such that its relationship with disorder risk is accounted for, at least in part, by common genetic effects. At this initial stage of inquiry, although several promising candidates have been identified, no single factor has been shown to satisfy all four of these criteria as they apply to alcoholism.

A major strategy used to identify possible mediating factors in alcoholism research is the high-risk design. Individuals with an alcoholic biological parent ("high-risk" offspring) are approximately 6 times more likely to develop alcoholism than are individuals whose biological parents are nonalcoholic ("low-risk" offspring). Moreover, twin and adoption studies indicate that this familial risk is due largely, but perhaps not entirely, to genetic factors. Consequently, factors that differentiate high-risk from low-risk offspring are candidate markers of inherited risk. In addition, if the offspring are assessed prior to their own exposure to alcohol, these candidate markers are more likely to be causes than consequences of the disorder, although this possibility needs to be assessed longitudinally, an assessment that has been made only rarely in alcoholism high-risk research. Four major domains have been implicated as possible inherited mediators in alcoholism by application of the high-risk design: alcohol sensitivity, comorbid psychopathology, personality, and neurophysiology.

Alcohol Sensitivity

Individuals differ widely in their psychological and pharmacological response to alcohol; experience with alcohol can be pleasurable and activating for some, but noxious and discomforting for others. Experimental research both with humans (Martin et al., 1985) and with other animals (McClearn & Kakihana, 1981) has demonstrated that genetic factors influence individual differences in alcohol sensitivity, providing support for the proposition that genetic influences on human drinking behavior might be mediated, in part, by inherited effects on alcohol sensitivity. Paradoxically, researchers have hypothesized that both hypersensitivity and hyposensitivity to alcohol may underlie inherited influences on alcoholism.

Finn and his colleagues (Finn and Pihl, 1987, 1988) have been the major proponents of the hypothesis that hypersensitivity to the effects of alcohol represents an inherited mediating factor in the etiology of alcoholism. There is a general consensus, among both researchers and clinicians, that the relief of psychological distress is a major motivating factor for drinking (i.e., the "self-medication" hypothesis). Finn and Pihl (1987, 1988) reported greater attenuation of the cardiovascular response to unavoidable shock in sons of alcoholics as compared to sons of non-alcoholics. Levenson, Oyama, and Meek (1987) reported similar findings when comparing daughters of alcoholics and nonalcoholics. Individuals at high risk for developing alcoholism may thus inherit increased sensitivity to the stress-response-dampening effects of alcohol.

Alternatively, Schuckit and his colleagues have been the major proponents of the hypothesis that hyposensitivity to the effects of alcohol represents an inherited risk factor in alcoholism. In a program of research that has extended over 15 years, Schuckit and his colleagues have reported that nonalcoholic, young adult male offspring of alcoholics consistently report reduced sensation of intoxication after a standard dose of alcohol as compared to the responses of nonalcoholic, young adult male offspring of nonalcoholics matched for drinking history (Schuckit & Gold, 1988). The interpretation of this result is clouded by the fact that while reduced sensitivity is consistently observed when subjectively assessed, objective measures (e.g., body sway, hormonal response) tend to produce inconsistent results (Sher, 1991). Schuckit has not articulated a comprehensive theoretical model that would account for an association between hyposensitivity to alcohol and alcoholism risk, although the underlying mechanism presumably involves alcohol overconsumption due either to lack of feedback inhibition or to the need to consume large amounts of alcohol to achieve desired psychological or pharmacological effects, or both. In an important extension of his work, Schuckit (1994) recently reported preliminary results from a 10-year follow-up of participants in the alcohol-challenge studies. He found that among men who showed diminished response to alcohol, 43% had developed alcoholism some time during the 10-year follow-up interval. The comparable rate of alcoholism among those with heightened sensitivity was only 11%.

In an important review article, Newlin and Thompson (1990) proposed a differentiator model to account for the apparent inconsistency of findings relating alcohol sensitivity to risk for alcoholism. In their review of the literature, they noted that high-risk individuals tend to show a heightened response to alcohol during rising blood alcohol concentration (BAC) (up to approximately 30 minutes following alcohol ingestion), but a diminished response during falling BAC. As alcohol's euphoric effects tend to be associated with rising BAC while its dysphoric effects tend to be associated with falling BAC, the differentiator model posits greater reinforcement of drinking among sons of alcoholics as compared to sons of nonalcoholics throughout the BAC curve. The differentiator model provides an integrative framework for future research in this area and has already gained some empirical support (Cohen, Porjesz, & Begleiter, 1993).

Comorbid Psychopathology

Recent epidemiological studies indicate that alcoholics are more likely than nonalcoholics to meet criteria for at least one other psychiatric diagnosis. Indeed, 44% of male and 65% of female alcoholics meet criteria for at least one additional psychiatric diagnosis (Helzer & Pryzbeck, 1988), with the prevalence of virtually every psychiatric disorder being higher among alcoholics as compared to nonalcoholics (Fig. 1). Given the substantial body of evidence that implicates genetic factors in the inheritance of most psychiatric disorders, it is possible that the inheritance of alcoholism in some cases may be secondary to the inheritance of other psychiatric disorders.

An association between alcoholism and another psychiatric disorder may reflect the effect of chronic alcohol abuse on psychiatric disease liability, the effect of psychiatric disorder on drinking behavior, or the effect of some third factor on the risks for both alcoholism and psychiatric disorder. One approach to deciding among these possibilities is to distinguish which disorder is primary and which is secondary by sequencing their onsets (e.g., Schuckit, 1985). Another approach has been to use family studies. For example, several studies have reported failure to find increased risk of alcoholism among the first-degree relatives of probands with a primary diagnosis of depression (e.g., Merikangas, Leckman, Prusoff, Pauls, & Weissman, (1985). These findings indicate that the familial component in depression is distinct from the familial component in alcoholism and therefore

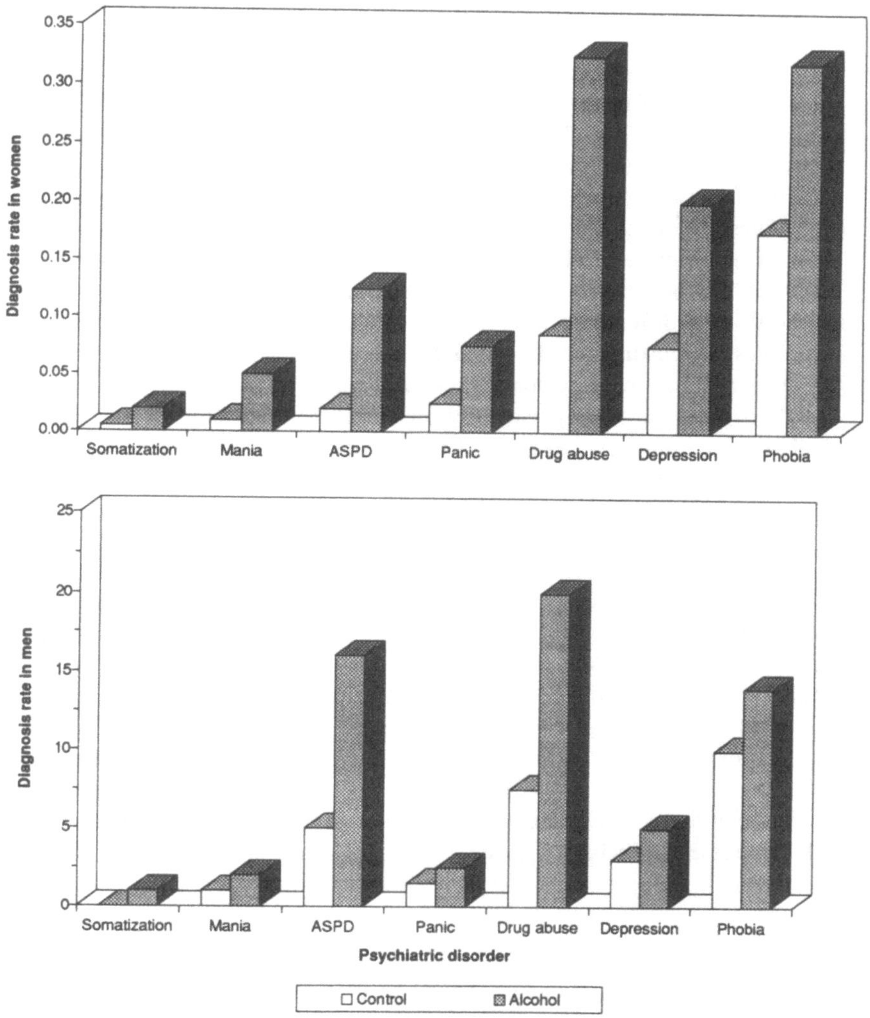

FIGURE 1. Rates of psychiatric disorder in alcoholics and nonalcoholics from a United States epidemiological study. Lifetime rates of diagnosis among women (A) and among men (B). Adapted from Helzer and Pryzbeck (1988) with permission.

suggest that the association between alcoholism and depression is most likely the result of the effect of chronic alcohol abuse on risk for depression.

Family study methods are only beginning to be used to resolve the nature of the associations between alcoholism and other psychiatric disorders. The interested reader is directed to recent reviews on the association between alcoholism and anxiety disorders (Kushner, Sher, & Beitman, 1990) and antisocial personality disorder (Sher and Trull, 1994) for additional information.

Personality Dimensions

Personality has long been implicated in alcoholism etiology. There is a substantial body of behavior genetic research that establishes a heritable component to personality (e.g., Bouchard, Lykken, McGue, Segal, & Tellegen, 1990). Cross-sectional studies have documented consistent, albeit modest, personality differences between alcoholics and controls (e.g. Barnes, 1983), while longitudinal studies have demonstrated that some of these personality differences predate alcoholism onset (e.g., Loper, Kammeier, & Hoffman, 1973). Significantly, many of the personality dimensions that differentiate alcoholics from nonalcoholics also differentiate the children of alcoholics from the children of nonalcoholics, suggesting that the inheritance of alcoholism is mediated, in part, by the inheritance of these personality characteristics.

Two dimensions of personality appear to be relevant to the etiology of alcoholism (for a review, see Sher, 1991). The first is negative emotionality, or the tendency to experience negative mood states and psychological distress (often referred to as "neuroticism"). As compared to the children of nonalcoholics, the children of alcoholics are more likely to be rated as neurotic and anxious and to be diagnosed with a depressive disorder. The second relevant dimension is behavioral control, or the willingness to inhibit behavior and endorse generally accepted social norms. As compared to the children of nonalcoholics, the children of alcoholics are more likely to be rated as inattentive and impulsive and to be diagnosed as conduct-disordered or as suffering from attention-deficit disorder.

These results suggest that developmental processes that end in adulthood with the expression of alcoholism can begin in preadolescence as personality deviations in negative emotionality or behavioral control or both. Understanding the nature of these developmental processes is a significant question that remains to be addressed by behavior geneticists in this area.

Neurophysiology

Although the initial interest in investigating neurophysiological markers in alcoholism was to document the neurological effects of chronic alcohol abuse, most current research is focused on their use as prospective biological markers of risk (Begleiter & Porjesz, 1990). Two neurophysiological markers have been associated with alcoholism risk in high-risk studies. The first, and most consistently associated, is amplitude of the P3 event-related brain potential (ERP). The P3 is a late positive deflection of the ERP wave-form that is thought to reflect allocation of attentional resources during memory update (Polich, Pollock, & Bloom, 1994). Begleiter, Porjesz, Bihari, and Kissin (1984) reported that preadolescent sons of alcoholic fathers showed significantly diminished P3 amplitude in a visual oddball paradigm as compared to the preadolescent sons of nonalcoholic fathers.

Since this initial report, several groups have tried to extend the findings of Begleiter and colleagues by investigating auditory as well as visual ERP and by sampling adolescent and young adult as well as preadolescent offspring of alcoholics. Although findings from these studies are not entirely consistent, a recent meta-analysis conduced that overall, males with a positive family history of alcoholism showed reduced P3 amplitude as compared to control males (Polich et al., 1994); regrettably, the small number of studies that included female offspring precludes a similar analysis of P3 amplitude in the daughters of alcoholics. Moreover, the reduction in P3 amplitude relative to controls was significantly larger for high-risk males aged 17 years and younger than for those aged 18 years and older. Although there currently is no articulated conceptual model that links alcoholism risk to P3 amplitude, the meta-analytical findings suggest the interesting possibility that reduced P3 amplitude indexes delayed neurological development in high-risk as compared to low-risk males.

The second neurophysiological marker associated with alcoholism risk is the EEG. The resting EEG of sober alcoholics evidences increased activity in the delta, theta, and beta ranges and decreased activity in the alpha range as compared to the resting EEG of nonalcoholics (Begleiter & Platz, 1972). Gabrielli et al. (1982) reported increased fast activity (i.e., > 17 Hz) in resting EEG in high-risk as compared to low-risk sons aged 11–13 years. While some have replicated the findings of Gabrielli and colleagues (e.g., Ehlers & Schuckit, 1990), most have not (for a review, see Cohen, Porjesz, & Begleiter, 1991), and the relevance of resting EEG in alcoholism etiology remains uncertain.

Somewhat more consistent are studies that relate alcoholism risk to EEG changes following alcohol ingestion. Thus, both Pollock et al. (1983) and Volavka, Pollock, Gabrielli, and Mednick (1985) reported greater increases in slow alpha energy following an alcohol challenge among high-risk as compared to low-risk subjects. Cohen et al. (1993) provided further support for differential alcohol-induced EEG changes in high-risk and low-risk males and interpreted their results as supportive of Newlin and Thomson's differentiator model. In particular, the Cohen et al. (1993) study suggests that high-risk individuals, as compared to low-risk individuals, are more likely to experience the relaxed and comfortable state that is associated with slow alpha activity during the rising BAC limb and thus to derive greater reinforcement value from their drinking.

Moderators of Alcoholism Inheritance

If the heritability of alcoholism varies reliably with status on another variable, that variable is said to be a moderator of inheritance. Two variables, age of alcoholism onset and gender, have been implicated as moderators of alcoholism inheritance.

Age of Onset

It has long been recognized that alcoholism is a heterogeneous disorder, and attempts to subclassify alcoholics phenotypically date back at least 100 years (Babor & Lauerman, 1986). Despite this long history of scientific and clinical attention, however, most alcoholism classification schemes have lacked replicability and predictive utility. Nonetheless, one classificatory distinction among alcoholics has consistently emerged and does appear to be etiologically significant. Some alcoholics are characterized by a relatively early age of problem-drinking onset, abuse of drugs other than alcohol, evidence of childhood aggression, hyperactivity and adult antisocial behavior, and poor clinical prognosis, while other alcoholics are characterized by a relatively late age of problem-drinking onset, fewer childhood behavioral risk factors and adult antisocial behavior, and relatively good clinical prognosis (Babor et al., 1992).

Evidence supporting the clinical significance of this classification comes from studies that show a consistent relationship between various indicators of clinical functioning and age of problem-drinking onset. Buydens-Branchey, Branchey and Noumair (1989) found that, when compared to those with a late onset of problem drinking, male alcoholics with an onset of problem drinking prior to age 20 years had a 50% higher rate of depression, were 4 times more likely to have attempted suicide, and were twice as likely to have been incarcerated for a violent crime. Similarly, Roy et al. (1991) reported that male alcoholics with an early age of onset were more likely than those with a late age of onset to have a lifetime diagnosis of hypomanic disorder or bipolar disorder and a history of suicide attempts. Clinical differences between early- and late-onset male alcoholics also appear to hold for female alcoholics. Glenn and Nixon (1991) reported that early-onset female alcoholics were more anxious, less socialized, and more likely to abuse cocaine than were late-onset female alcoholics.

Cloninger (1987) was one of the first to emphasize the etiological significance of age of onset in his Type I/Type II model of alcoholism. Although the Cloninger typological model has failed to gain consistent empirical support (e.g., Schuckit and Irwin, 1989), resulting in the introduction of empirically based classifications (e.g., Babor et al., 1992), Cloninger's specific hypothesis that inherited factors are more influential in early-onset as compared to late-onset alcoholics has been supported. Buydens-Branchey et al. (1989), Roy et al. (1991), Glenn and Nixon (1991), and Penick et al. (1987) all reported that early-onset alcoholics are more likely than late-onset alcoholics to have a positive family history of alcoholism. In our study of male alcoholic twins, we reported findings indicating that the greater familial loading of early- as compared to late-onset alcoholism was due to genetic and not environmental effects (McGue, Pickens, & Svkis, 1992); a similar analysis with female twins did not produce significant differences. Among male alcoholics with an onset of problem drinking prior to age 20 years, concordance (i.e., risk to co-twin of an affected proband) was 86.5% for MZ but only 56.8% for DZ twins. In contrast, among male alcoholics with an onset after age 20 years, concordance was 60.6% in MZ and 50.9% in DZ twins. Heritability estimates derived from a biometrical analysis of the twin concordances further substantiated the existence of differential genetic influence; heritability was significantly higher among early- as compared

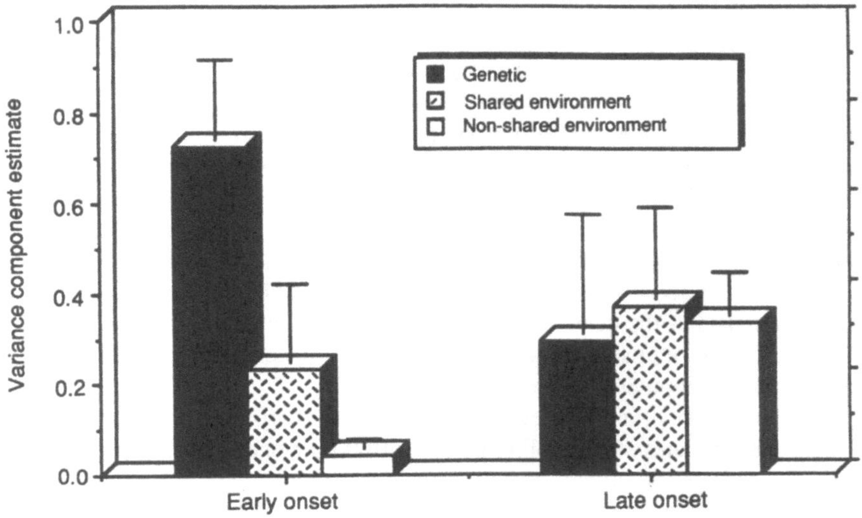

FIGURE 2. Decomposition of alcoholism liability variance from a sample of male twins classified according to proband onset of problem drinking being either before (Early Onset) or after (Late Onset) age 20 years. The figure gives the proportion of liability variance associated with genetic, shared environmental, and nonshared environmental factors. The heritability is significantly greater among early- as compared to late-onset pairs. Adapted from McGue et al. (1992) with permission.

to late-onset alcoholics (Fig. 2). The environmental mechanisms that underlie reduced heritability in late-onset alcoholism remain to be identified.

Gender

The lifetime morbid risk of alcoholism is estimated to be 2–4 times higher among males than among females (Helzer, Burnam, & McEvoy, 1991). Behavior genetic methodology can help identify the mechanisms that underlie this sex difference in prevalence. Two questions are relevant. First, is the strength of genetic influence (i.e., the heritability) the same in men and women? Second, are the genes involved in the etiology of alcoholism the same for men and women? These questions are considered further in Chapter 10. Unfortunately, despite the significance of these questions, available research does not allow their unequivocal resolution.

Sex Differences in Heritability. Relevant twin and adoption studies (reviewed above) indicate that (1) MZ twin concordance for alcoholism is substantially higher among men than among women; (2) higher MZ than DZ concordance for alcoholism has been observed consistently in studies of men, but inconsistently in studies of women; and (3) increased risk of alcoholism has been observed consistently among the reared-away sons of alcoholics, but inconsistently among the reared-away daughters of alcoholics. This pattern of results led us to hypothesize that the heritability of alcoholism is lower among women as compared to men (McGue et al., 1992; McGue, 1994).

The heritability of the liability for developing alcoholism (rather than alcoholism per se) can be derived from twin studies under the assumption that alcoholism is a threshold character (Fig. 3) (Falconer, 1965). That is, we assume that underlying the categorical diagnosis of alcoholism is an unobserved but normally distributed liability such that the disorder is expressed whenever an individual's combined liability exceeds some threshold value along the liability continuum. Under this assumption, the inheritance of alcoholism is secondary to the inheritance of the quantitative liability, and the heritability of the underlying liability can be estimated from the observed twin concordance rates using recent analytical developments (Rice & Reich, 1985).

Application of these methods in twin studies of male alcoholics yields consistent estimates of heritability in the 0.40–0.60 range, with the single exception of a heritability estimate of 0.90 derived from concordances reported by Kaij (1960) in a small sample of Swedish twins (McGue, 1994). In contrast, heri-

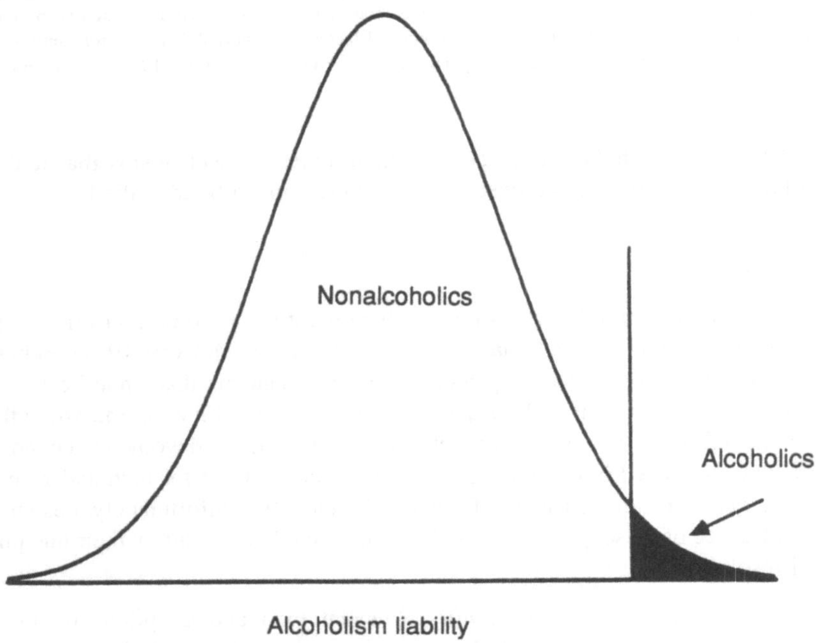

FIGURE 3. Liability-threshold model as applied to alcoholism. It is assumed that underlying the categorical distinction between alcoholics and nonalcoholics is a continuously distributed liability. Alcoholism occurs whenever an individual's combined liability exceeds a fixed threshold value along the liability continuum. According to this model, the inheritance of alcoholism is secondary to the inheritance of the underlying liability.

tability estimates from twin studies of female alcoholics are rather inconsistent and cluster around two values, 0.0 (i.e., no heritability) and 0.50 (i.e., a degree of heritability comparable to that observed in men) (McGue, 1994). Although it might be tempting to average these estimates to produce an intermediate value less than the heritability estimates observed in men, aggregation of numbers as disparate as these is potentially misleading when one does not know the source of the inconsistency. Indeed, in a recent meta-analysis, Heath et al. (1994) demonstrated that when differences in male and female alcoholism prevalence rates are taken into account, unambiguous evidence of a difference in heritability between men and women is still lacking. Consequently, while the available evidence does allow one to conclude that alcoholism liability is moderately heritable in men, the specific degree to which it is heritable in women remains uncertain.

Cross-Sex Transmission. Family studies of male and female alcoholics can help to identify elements common to the etiology of alcoholism in the two sexes. In particular, the finding of increased risk of alcoholism among the opposite-sex relatives of alcoholic index cases (i.e., cross-sex transmission) would indicate shared familial transmission. In a recent meta-analysis of family studies that investigated risk of alcoholism in both the male and female relatives of alcoholics, McGue and Slutske (1995) reported that the cross-sex and within-sex transmission rates were comparable. That is, the pooled (across-study) rate of alcoholism among the male first-degree relatives of alcoholics was 0.266 if the index case was male and 0.334 if the index case was female. The comparable pooled rates of alcoholism among female first-degree relatives were 0.060 and 0.103, respectively. In a large study published after the McGue and Slutske review was written, Kendler et al. (1994) reported similar rates of alcoholism among the daughters of male and female alcoholics, providing further support for the general conclusion that within-sex and between-sex transmission rates are similar.

The proposition that alcoholism liability is less heritable in women than in men appears to be at odds with the conclusion that the cross-sex and within-sex transmission rates are comparable. There are several admittedly ad hoc and conflicting explanations for this apparent inconsistency. First, the hypothesis of reduced heritability in females may be invalid, as several workers have suggested (e.g., Kendler et al., 1992; Heath et al., 1994). Indeed, as discussed above, the available behavior genetic literature does not permit any firm conclusions about the heritability of liability to alcoholism in women. Alternatively, if the heritability of alcoholism liability is reduced in women, the effect of this reduced heritability on familial transmission rates might be compensated for by a multiple threshold effect. That is, for many genetically influenced disorders in which there is a sex difference in prevalence, familial risk is higher among the relatives of index cases from the less frequently, as compared to the more frequently, affected sex. Thus, the rate of mental retardation is higher among males than among females, but the risk to relatives is higher if the index case is female rather than male (Pauls, 1979).

A similar pattern is observed with pyloric stenosis (Emery & Rimoin, 1983) and antisocial personality disorder (Cloninger, Christiansen, Reich, & Gottesman, 1978). The standard explanation for this distinctive pattern is that, because members of the less frequently affected sex are buffered against the expression of the disorder, they require a higher loading of predisposing genetic factors before the disorder becomes manifest. This higher genetic loading is then passed on to relatives as higher risk for the disorder (Cloninger et al., 1978). In the case of alcoholism, failure to find increased risk of alcoholism among the relatives of female, as compared to male, index cases may be evidence of reduced heritability in women. It is evident that the question of gender moderation of alcoholism inheritance remains an important yet clearly unresolved question in the alcohol research field.

SUMMARY

Adoption, twin, and genetic marker studies aimed at identifying the existence of genetic influences on risk for alcoholism were reviewed. Findings from these studies converge on the conclusion that there are genetic influences on risk for alcoholism. These studies also serve to confirm the importance of environmental factors in alcoholism etiology: (1) Identical twin concordance for alcoholism is consistently substantially less than 1 (especially among women); (2) individuals reared with alcoholics appear to carry an excess risk of alcoholism even when the basis of their familial relationship is adoptive and not biological; and (3) some individuals who inherit genes that are normally protective against drinking can apparently develop alcoholism if they exist within a culture that encourages and reinforces high-density alcohol consumption.

Although additional twin and adoption studies might help increase our confidence in concluding that genes influence alcoholism, it was argued that the field is now appropriately more interested in attempting to determine how, rather than simply whether, genes influence risk for alcoholism. Methods that might be brought to bear on this question are discussed further in Chapters 12 and 13. Molecular genetic research in alcoholism should ultimately lead to the discovery of the specific genes that contribute to alcoholism liability and thus initiate the process of mapping the complex pathway that links genes to behavior by starting at the primary gene product level. Additionally, because progress in molecular genetic mapping of complex behavioral phenotypes such as alcoholism is likely to be slow and uncertain, there is a need for a complementary, top-down, strategy for mapping the gene-to-behavior pathway by starting at the behavioral level and identifying intermediate behavioral, neurophysiological, metabolic, and neurochemical factors. Current research that adopts this top-down strategy was characterized in terms of attempting to identify mediators and moderators of alcoholism inheritance. Examples of both types of factors were given.

One criticism of behavior genetic research on alcoholism is its atheoretical, empirical nature. Although behavior geneticists have taught us much about the empirical correlates of alcoholism, there is as yet no comprehensive conceptual model that specifies the mechanism of genetic influence. The development of comprehensive theoretical models of alcoholism etiology awaits integration of behavior genetic research with the substantial body of research linking environmental factors to alcoholism risk. For example, alcoholism rates vary with ethnicity, across cohorts, over the life span, according to religious convictions, and as a function of access of alcohol [e.g., bar hours (Smith, 1980)]. Alcoholism is thus unique among the psychiatric phenotypes investigated by behavior geneticists in that its expression appears to be highly dependent on cultural factors. It would seem that understanding the mechanism of genetic influence on alcoholism will ultimately require integrative approaches that incorporate both biological and psychosocial risk.

ACKNOWLEDGMENTS. The research reported in this chapter was supported in part by Grants R01-DA05147, RO1-AA09367, and KO2-AA00175.

REFERENCES

Abrams, D. B., & Niaura, R. S. (1987). Social learning theory. In H. T. Blane & K. E. Leonard (Eds.), *Psychological theories of drinking and alcoholism* (pp. 131–178). New York: Guilford Press.

Agarwal, D. P., & Goedde, H. W. (1989). Human aldehyde dehydrogenases: Their role in alcoholism. *Alcohol, 6,* 517–523.

Anastasi, A. (1958). Heredity, environment, and the question "how?" *Psychological Bulletin, 65,* 197–208.

Babor, T. F., Hofmann, M., DelBoca, F. K., Hesslebrock, V., Meyer, R. E., Dolinsky, Z. S., & Rounsaville, B. (1992). Types of alcoholics. I. Evidence for an empirically derived typology based on indicators of vulnerability and severity. *Archives of General Psychiatry, 49,* 599–608.

Babor, T. F., & Lauerman, R. J. (1986). Classification and forms of inebriety: Historical antecedents of alcoholic typologies. In M. Galanter (Ed.), *Recent advances in alcoholism,* Vol. 4 (pp. 113–144). New York: Plenum Press.

Barnes, G. E. (1983). Clinical and prealcoholic personality characteristics. In B. Kissin & H. Begleiter (Eds.), *The pathogenesis of alcoholism: Psychosocial issues.* (pp. 113–181). New York: Plenum Press.

Baron, R. M., & Kenny, D. A. (1986). The moderator–mediator variable distinction in social psychological research: Conceptual, strategic, and statistical considerations. *Journal of Personality and Social Psychology, 51,* 1173–1182.

Begleiter, H., & Platz, A. (1972). The effects of alcohol on the central nervous system in humans. In B. Kissin & H. Begleiter (Eds.), *The biology of alcoholism,* Vol. 2, *Physiology and behavior* (pp. 293–306). New York: Plenum Press.

Begleiter, H., & Porjesz, B. (1990). Neuroelectric processes in individuals at risk for alcoholism. *Alcohol and Alcoholism, 25,* 251–256.

Begleiter, H., Porjesz, B., Bihari, B., & Kissin, B. (1984). Event-related brain potential in boys at risk for alcoholism. *Science, 225,* 1493–1496.

Blum, K., Noble, E. P., Sheridan, P. J., Montgomery, A., Ritchie, T., Jagadeeswaran, P., Nogam, H., Briggs, A. H., & Cohen, J. B. (1990). Allelic association of human dopamine D2 receptor gene and alcoholism. *Journal of the American Medical Association, 263,* 2055–2060.

Bohman, M., Sigvardsson, S., & Cloninger, C. R. (1981). Maternal inheritance of alcohol abuse: Cross-fostering analysis of adopted women. *Archives of General Psychiatry, 38*, 965–969.

Bolos, A. M., Dean, M., Lucas-Derse, S., Ramsburg, M., Brown, G. L., & Goldman, D. (1990). Population and pedigree studies reveal a lack of association between the dopamine D2 receptor gene and alcoholism. *Journal of the American Medical Association, 264*, 3156–3160.

Botstein, D., White, R. L., Skolnick, M., & Davis, R. W. (1980). Construction of a genetic linkage map in man using restriction fragment length polymorphisms. *American Journal of Human Genetics, 32*, 314–331.

Bouchard, T. J., Lykken, D. T., McGue, M., Segal, N. L., & Tellegen, A. (1990). Sources of human psychological differences: The Minnesota Study of Twins Reared Apart. *Science, 250*, 223–228.

Buydens-Branchey, L., Branchey, N. H., & Noumair, D. (1989). Age of alcoholism onset: Relationship to psychopathology. *Archives of General Psychiatry, 46*, 225–230.

Cadoret, R. J., O'Gorman, T., Troughton, E., & Heywood, E. (1985). Alcoholism and antisocial personality: Interrelationships, genetic and environmental factors. *Archives of General Psychiatry, 42*, 161–167.

Cadoret, R. J., Troughton, E., & O'Gorman, T. W. (1987). Genetic and environmental factors in alcohol abuse and antisocial personality. *Journal of Studies on Alcohol, 48*, 1–8.

Caldwell, C. B., & Gottesman, I. I. (1991). Sex differences in risk for alcoholism: A twin study. *Behavior Genetics, 21*, 563–563 (abstract).

Cloninger, C. R. (1987). Neurogenetic adaptive mechanisms in alcoholism. *Science, 236*, 410–416.

Cloninger, C. R., Bohman, M., & Sigvardsson, S. (1981). Inheritance of alcohol abuse: Cross-fostering analysis of adopted men. *Archives of General Psychiatry, 38*, 861–868.

Cloninger, C. R., Christiansen, K. O., Reich, T., & Gottesman, I. I. (1978). Implications of sex differences in the prevalence of antisocial personality, alcoholism, and criminality for familial transmission. *Archives of General Psychiatry, 35*, 941–951.

Cohen, H. L., Porjesz, B., & Begleiter, H. (1991). EEG characteristics in males at risk for alcoholism. *Alcoholism: Clinical and Experimental Research, 15*, 858–865.

Cohen, H. L. Porjesz, B., & Begleiter, H. (1993). The effects of ethanol on EEG activity in males at risk for alcoholism. *Electroencephalography and Clinical Neurophysiology, 86*, 368–376.

Comings, D. E., Comings, B. G., Muhleman, D., Dietz, G., Shahbahrumi, B., Tast, D., Knell, E., Kocsis, P., Baumgarten, R., Kovacs, B. W., Levy, D. L., Smith, M., Borison, R. L., Evans, D. D., Klein, D. N., MacMurray, J., Tosk, J. M., Sverd, J., Gysin, R., & Flanagan, D. D. (1991). The dopamine D2 receptor locus as a modifying gene in neuropsychiatric disorders. *Journal of the American Medical Association, 266*, 1793–1800.

Conneally, P. M. (1991). Association between D2 dopamine receptor gene and alcoholism: Continuing controversy. *Archives of General Psychiatry, 48*, 757–759.

Cook, B. L., Wang, Z. W., Crowe, R. R., Hauser, R., & Freimer, M. (1992). Alcoholism and the D2 receptor gene. *Alcoholism: Clinical and Experimental Research, 16*, 806–809.

Cotton, N. S. (1979). The familial incidence of alcoholism: A review. *Journal of Studies on Alcohol, 40*, 89–116.

Crabb, D. W., Edenberg, H. J., & Borson, W. F. (1987). Genotypes for aldehyde dehydrogenase deficiency and alcohol sensitivity: The inactive ALDH2 allele is dominant. *Journal of Clinical Investigation, 83* 314–316.

Egeland, J. A., Gerhard, D. S., Pauls, D. S., Sussex, J. N., & Kidd, K. K. (1987). Bipolar affective disorders linked to DNA markers on chromosome 11. *Nature, 325*, 783–787.

Ehlers, C. L., & Schuckit, M. A. (1990). EEG fast frequency in the sons of alcoholics. *Biological Psychiatry, 27*, 631–641.

Emery, A. E. H., & Rimoin, D. L. (1983). *Principles and practice of medical genetics*. Edinburgh: Churchill Livingston.

Falconer, D. S. (1965). The inheritance of liability to certain diseases estimated from the incidence among relatives. *Annals of Human Genetics, 29*, 51–76.

Finn, P. R., & Pihl, R. O. (1987). Men at high risk for alcoholism: The effect of alcohol on cardiovascular response to unavoidable shock. *Journal of Abnormal Psychology, 96,* 230–236.

Finn, P. R., & Pihl, R. O. (1988). Risk for alcoholism: A comparison between two different groups of sons of alcoholics on cardiovascular reactivity and sensitivity to alcohol. *Alcoholism: Clinical and Experimental Research, 12,* 742–747.

Gabrielli, W. F., Mednick, S. A., Volavka, J., Pollock, V. E., Schulsinger, F., & Itil, T. M. (1982). Electroencephalograms in children of alcoholic fathers. *Psychophysiology, 19,* 404–407.

Gelernter, J., Goldman, D., & Risch, N. (1993). The A1 allele at the D_2 dopamine receptor gene and alcoholism: A reappraisal. *Journal of the American Medical Association, 269,* 1673–1677.

Gelernter, J., O'Malley, S., Risch, N., Kramzler, H. R., Krystal, J., Merikangas, K., Kennedy, J. L., & Kidd, K. K. (1991). No association between an allele at the D2 dopamine receptor gene (DRD2) and alcoholism. *Journal of the American Medical Association, 266,* 1801–1807.

Glenn, S. W., & Nixon, S. J. (1991). Applications of Cloninger's subtypes in a female alcoholic sample. *Alcoholism: Clinical and Experimental Research, 5,* 851–857.

Goedde, H. W., Agarwal, D. P., & Fritze, G. (1992). Distribution of ADH2 and ALDH2 genotypes in different populations. *Human Genetics, 88,* 344–346.

Goedde, H. W., Harada, S., & Agarwal, D. P. (1979). Racial differences in alcohol sensitivity: A new hypothesis. *Human Genetics, 51,* 331–334.

Goldman, D., Dean, M., Brown, G. L., Bolos, A. M., Tokola, R., Virkkunen, M., & Linnoila, M. (1992). D2 dopamine receptor genotype and cerebrospinal fluid homovanillic acid, 5-hydroxyindoleacetic acid and 3-methoxy-4-hydroxyphenylglycol in Finland and the United States. *Acta Psychiatrica Scandinavica, 86,* 351–357.

Goodwin, D. W., Schulsinger, F., Hermansen, K., Guze, S. B., & Winokur, G. (1973). Alcohol problems in adoptees raised apart from alcoholic biological parents. *Archives of General Psychiatry, 28,* 238–243.

Goodwin, D. W., Schulsinger, F., Knop, J., Mednick, S., & Guze, S. B. (1977). Alcoholism and depression in the adopted-out daughters of alcoholics. *Archives of General Psychiatry, 34,* 751–755.

Gurling, H. M. D., Oppenheim, B. E., & Murray, R. M. (1984). Depression, criminality and psychopathology associated with alcoholism: Evidence from a twin study. *Acta Geneticae Medicae et Gemellologiae, 33,* 333–339.

Harada, S., Agarwal, D. P., Goedde, H. W., Tagaki, S., & Ishikawa, B. (1982). Possible protective role against alcoholism for aldehyde dehydrogenase isozyme deficiency in Japan. *Lancet, 2,* 827–827.

Heath, A. C., Jardine, R., & Martin, N. G. (1989). Interactive effects of genotype and social environment on alcohol consumption in female twins. *Journal of Studies on Alcohol, 60,* 38–48.

Heath, A. C., Slutske, W. S., & Madden, P. A. F. (1995). Gender differences in the genetic contribution to alcoholism risk and drinking patterns. In R. W. Wilsnack & S. C. Wilsnack (Eds.), *Gender and alcohol.* Rutgers, NJ: Rutgers University Press.

Helzer, J. E., Burnam, A., & McEvoy, L. T. (1991). Alcohol abuse and dependence. In L. N. Robins & D. A. Reiger (Eds.), *Psychiatric disorders in America: The Epidemiological Catchment Area study* (pp. 81–115). New York: Free Press.

Helzer, J., Canino, G., & Yeh, E. K. (1990). Alcoholism—North America and Asia. *Archives of General Psychiatry, 47,* 313–319.

Helzer, J. E., & Pryzbeck, T. R. (1988). The co-occurrence of alcoholism with other psychiatric disorders in the general population and its impact on treatment. *Journal of Studies on Alcohol, 49,* 219–224.

Hrubec, Z., & Omenn, G. S. (1981). Evidence of genetic predisposition to alcoholic cirrhosis and psychosis: Twin concordances for alcoholism and its biological endpoints by zygosity among male veterans. *Alcoholism: Clinical and Experimental Research, 5,* 207–212.

Kaij, L. (1960). *Alcoholism in twins.* Stockholm: Almqvist & Wiksell.

Kendler, K. S., Heath, A. S., Neale, M. C., Kessler, R. C., & Eaves, L. J. (1992). A population based twin study of alcoholism in women. *Journal of the American Medical Association, 268,* 1877–1882.

Kendler, K. S., Neale, M. C., Keisler, R. C., Heath, A. C., and Eaves, L. J. (1993). A test of the equal-environment assumption in twin studies of psychiatric illness. *Behavior Genetics, 23,* 21–28.

Kushner, M., Sher, K. J., & Beitman, B. (1990). The relation between alcohol problems and the anxiety disorders. *American Journal of Psychiatry, 147,* 685–695.

Levenson, R. W., Oyama, O. N., & Meek, P. S. (1987). Greater reinforcement from alcohol for those at risk: Parental risk, personality risk, and sex. *Journal of Abnormal Psychology, 96,* 242–253.

Loehlin, J. C., & Nicholas, R. C. (1976). *Heredity, environment, and personality.* Austin: University of Texas Press.

Loper, R. G., Kammeier, M. L., & Hoffmann, H. (1973). MMPI characteristics of college freshman males who later became alcoholics. *Journal of Abnormal Psychology, 82,* 159–162.

Lytton, H., Martin, N. G., & Eaves, L. J. (1977). Environmental and genetical causes of variation in ethnological aspects of behavior in two-year-old boys. *Social Biology, 24,* 200–211.

Martin, N. G., Oakeshott, J. G., Gibson, J. B., Starmer, G. A., Perl, J., & Wilks, A. V. (1985). A twin study of psychomoter and physiological responses to an acute dose of alcohol. *Behavior Genetics, 15,* 305–347.

McClearn, G. E., & Kakihana, R. (1981). Selective breeding for ethanol sensitivity: Short-sleep versus long-sleep mice. In G. E. McClearn, R. A. Deitrich, & V. G. Erwin (Eds.), *Development of animal models as pharmacologic tools, National Institute on Alcoholism and Alcohol Abuse Research Monograph No. 60* (pp. 147–159). Rockville, MD: National Institute on Alcoholism and Alcohol Abuse.

McGue, M. (1992). When assessing twin concordance, use the probandwise not the pairwise rate. *Schizophrenia Bulletin, 18,* 171–176.

McGue, M. (1993). From proteins to cognitions: The behavioral genetics of alcoholism. In R. Plomin and G. E. McClearn (Eds.), *Nature, nurture and psychology* (pp. 245–268). Washington, DC: American Psychological Association.

McGue, M. (1994). Genes, environment and the etiology of alcoholism. In R. Zucker, G. Boyd, and J. Howard (Eds.), *Development of alcohol-related problems: Exploring the biopsychosocial matrix of risk. National Institute on Alcoholism and Alcohol Abuse Research Monograph No. 26.* Rockville MD: National Institute on Alcoholism and Alcohol Abuse, pp. 1–40.

McGue, M., Pickens, R. W., & Svikis, D. S. (1992). Sex and age effects on the inheritance of alcohol problems: A twin study. *Journal of Abnormal Psychology, 101,* 3–17.

McGue, M., & Slutske, W. (1995). *The inheritance of alcoholism in women, National Institute on Alcoholism and Alcohol Abuse Research Monograph.* Rockville MD: National Institute on Alcoholism and Alcohol Abuse.

McGuffin, P., Sargeant, M., Hetti, A., Tidmarsh, S., Whatley, S., & Marchbanks, R. M. (1990). Exclusion of a schizophrenia susceptibility gene from the chromosome 5q11–q13 region: New data and a reanalysis of previous reports. *American Journal of Human Genetics, 48,* 1206–1209.

Merikangas, K. R. (1990). The genetic epidemiology of alcoholism. *Psychological Medicine, 20,* 11–22.

Merikangas, K. R., Leckman, J. F., Prusoff, B. A., Pauls, D. L., & Weissman, M. M. (1985). Familial transmission of depression and alcoholism. *Archives of General Psychiatry, 42,* 367–372.

Murray, R. M., Clifford, C. A., & Gurling, H. M. D. (1983). Twin and adoption studies: How good is the evidence for a genetic role? In M. Galanter (Ed.), *Recent developments in alcoholism,* Vol. 1, *Genetics, behavioral treatment, social mediators, and prevention* (pp. 25–48). New York: Plenum Press.

Neale, M. C., & Cardon, L. C. (1992). *Methodology for genetic studies of twins and families*. Norwell, MA: Kluwer Academic Publishers.

Newlin, D. B., & Thompson, J. B. (1990). Alcohol challenge with sons of alcoholics: A critical review and analysis. *Psychological Bulletin, 108*, 383–402.

Parsian, A., Todd, R. D., Devor, E. J., O'Malley, K. L., Suarez, B. K., Reich, T., & Cloninger, C. R. (1991). Alcoholism and alleles of the human D2 dopamine receptor locus: Studies of association and linkage. *Archives of General Psychiatry, 48*, 655–663.

Pauls, D. L. (1979). Sex effect on risk of mental retardation. *Behavior Genetics, 9*, 289–296.

Peele, S. (1986). The implications and limitations of genetic models of alcoholism and other addictions. *Journal of Studies on Alcohol, 47*, 63–73.

Penick, E. C., Powell, B. J., Bingham, S. F., Liskow, B. I., Miller, N. S., & Read, M. R. (1987). A comparative study of familial alcoholism. *Journal of Studies on Alcohol, 48*, 139–146.

Pickens, R. W., Svikis, D. S., McGue, M., Lykken, D. T., Heston, L. L., & Clayton, P. J. (1991). Heterogeneity in the inheritance of alcoholism: A study of male and female twins. *Archives of General Psychiatry, 48*, 19–28.

Plomin, R., DeFries, J. C., & McClearn, G. E. (1990). *Behavioral genetics: A primer*, 2nd ed. San Francisco: W. H. Freeman.

Polich, J., Pollock, V. E., & Bloom, F. E. (1994). Meta-analysis of P300 amplitude from males at risk for alcoholism. *Psychological Bulletin, 115*, 55–73.

Pollock, V. E., Volavka, J., Goodwin, D. W., Mednick, S. S., Gabrielli, W. F., Knop, J., & Schulsinger, F. (1983). The EEG after alcohol in men at risk for alcoholism. *Archives of General Psychiatry, 40*, 857–861.

Rice, J., & Reich, T. (1985). Familial analysis of qualitative traits under multifactorial inheritance. *Genetic Epidemiology, 2*, 301–315.

Risch, N. (1990). Genetic linkage and complex diseases with special reference to psychiatric disorders. *Genetic Epidemiology, 7*, 3–16.

Roe, A. (1944). The adult adjustment of children of alcoholic parents raised in foster homes. *Quarterly Journal of Studies on Alcohol, 5*, 378–393.

Room, R. (1983). Region and urbanization as factors in drinking practices and problems. In B. Kissin & H. Begleiter (Eds.), *The pathogenesis of alcoholism: Psychological factors*, (pp. 555–604). New York: Plenum Press.

Roy, A., DeJong, J., Lamparski, D., Adinoff, B., George, T., Moore, V., Garnett, D., Kerich, M., & Linnoila, M. (1991). Mental disorders among alcoholics: Relationship to age of onset and cerebrospinal fluid neuropeptides. *Archives of General Psychiatry, 48*, 423–427.

Schuckit, M. A. (1985). The clinical implications of primary diagnostic groups among alcoholics. *Archives of General Psychiatry, 42*, 1043–1049.

Schuckit, M. A. (1994). Low level of response to alcohol as a predictor of future alcoholism. *American Journal of Psychiatry, 151*, 184–189.

Schuckit, M. A., & Gold, E. O. (1988). Simultaneous evaluation of multiple markers of ethanol/placebo challenges in sons of alcoholics and controls. *Archives of General Psychiatry, 45*, 211–216.

Schuckit, M. A., & Irwin, M. (1989). An analysis of the clinical relevance of Type 1 and Type 2 alcoholics. *British Journal of Addiction, 84*, 869–876.

Schwab, S., Soyka, M., Niederecker, M., Ackenheil, M., Scherer, J., & Wildenauer, D. B. (1991). Allelic association of human D2-receptor DNA polymorphism ruled out in 45 alcoholics. *American Journal of Human Genetics, Supplement, 49*, 203 (abstract).

Searles, J. S. (1988). The role of genetics in the pathogenesis of alcoholism. *Journal of Abnormal Psychology, 97*, 153–167.

Sher, K. J. (1991). *Children of alcoholics: A critical appraisal of theory and research*. Chicago: University of Chicago Press.

Sher, K. J., & Trull, T. J. (1994). Personality and disinhibitory psychopathology: Alcoholism and antisocial personality disorder. *Journal of Abnormal Psychology, 103*, 92–102.

Sherrington, R., Brynjolfsson, J., Petursson, H., Dudleston, K., Barraclough, B., Wasmuth, J., Dobbs, M., & Gurling, H. M. D. (1988). Localization of a susceptibility locus for schizophrenia on chromosome 5. *Nature, 336,* 164–167.

Smith, D. I. (1980). The introduction of Sunday alcohol sales in Perth: Some methodological observations. *Community Health Studies, 4,* 289–293.

Turner, E., Ewing, J., Shilling, P., Smith, T. L., Irwin, M., Schuckit, M., & Kelsoe, J. R. (1992). Lack of association between an RFLP near the dopamine D2 receptor gene and severe alcoholism. *Biological Psychiatry, 31,* 285–290.

Uhl, G., Blum, K., Noble, E., & Smith, S. (1993). Substance abuse vulnerability and D2 receptor genes. *Trends in Neuroscience, 16,* 83–88.

Vandenberg, S. G. (1976). Twin studies. In A. R. Kaplan (Ed.), *Human behavior genetics* (pp. 90–150). Springfield, IL: Charles J. Thomas.

Vogel, F., & Motulsky, A. G. (1986). *Human genetics: Problems and approaches.* New York: Springer-Verlag.

Volavka, J., Pollock, V., Gabrielli, W. F., & Mednick, S. A. (1985). The EEG in persons at risk for alcoholism. In M. Galanter (Ed.), *Recent developments in alcoholism* (pp. 21–36). New York: Plenum Press.

Wise, R. A., & Rompre, P. P. (1989). Brain dopamine and reward. *Annual Review of Psychology, 40,* 191–225.

CHAPTER THREE

Genetic Influences on Smoking Behavior

ANDREW C. HEATH AND PAMELA A. F. MADDEN

INTRODUCTION

In this chapter, we review evidence that suggests important genetic influences on smoking. These influences may affect the probability that an individual will become a smoker and, once he or she has started smoking, may affect how heavily the individual smokes and the probability that he or she will not quit smoking. We shall broadly summarize these measures under the heading "smoking behavior." In contrast to the many twin and adoption studies designed to explore the possibility of a genetic contribution to alcoholism risk (see Heath, Slutske, & Madden [1995] and Chapter 2), there has been comparatively little research focused on understanding the genetic contribution to smoking behavior (for an early exception to this statement, see Eaves & Eysenck, 1980). This neglect is surprising, when one considers that the adverse health consequences of smoking, both to the smoker (USDHEW, 1979) and to others in the smoker's environment (USDHHS, 1986), and the consequent enormous economic costs of smoking to society, are well established.

By some estimates, as many as 400,000 lives are lost each year to smoking in the United States alone (USDHHS, 1989; Peto, Lopez, Boreham, Thun, & Heath, 1992). A recent report of the Unites States Surgeon General noted that nicotine is a highly addictive substance (USDHHS, 1988) and indeed is more addictive than most illicit drugs. Simple prevention measures (e.g., the increase

ANDREW C. HEATH AND PAMELA A. F. MADDEN • Department of Psychiatry, Washington University School of Medicine, St. Louis, Missouri 63110.

Behavior Genetic Approaches in Behavioral Medicine, edited by J. Rick Turner, Lon R. Cardon, and John K. Hewitt. Plenum Press, New York, 1995.

in taxation of cigarettes in Canada in the early 1980s) can greatly decrease the numbers of individuals who become smokers. However, exclusive focus on prevention of smoking *initiation* ignores the enormous pool of continuing smokers [as many as 40 million in the United States alone (USDHHS, 1988)] and the future epidemic of health problems that is to be anticipated if those individuals who are current smokers do not quit. We will argue that behavior genetic studies can make an important contribution to understanding why these individuals persist in smoking.

Although there have been few if any behavior genetic studies designed with the primary goal of understanding the genetic influence on smoking behavior, available data do support an important genetic influence. Indeed, considering the imprecision and variability of estimates of the genetic contribution to alcoholism risk (Heath et al., 1995), it is arguable that the case for a substantial genetic influence on smoking behavior is even stronger than the case for such an influence on alcoholism. Much of the available data derives from large epidemiological surveys of twin panels in Scandinavia (principally Sweden and Finland) and the United States (male veteran twins from World War II), conducted from the 1960s onward. In these studies, the twin design was used to provide a more sensitive test of the adverse health consequences of smoking (by comparing smokers and their nonsmoking co-twins), and smoking was assessed as a key risk factor, rather than as a behavioral phenotype whose mode of inheritance was to be investigated (e.g., Raaschou-Nielsen, 1960; Cederlof, Epstein, Friberg, Hrubec, & Radford, 1971; Medlund, Cederlof, Floderus-Myrhed, Friberg, & Sorensen, 1977; Hrubec & Neel, 1978; Kaprio, Sarna, Koskenvuo, & Rantasalo, 1978; Carmelli, Swan, Robinette, & Fabsitz, 1992).

We shall draw on published data from these studies, as well as on our own data from studies of the Virginia twin panel (Heath et al., 1993a; Kendler et al., 1993) and the Australian National Health and Medical Research Council (NH & MRC) volunteer twin panel (Heath & Martin, 1993; Heath, 1993a; Madden et al., 1993a–c). These latter data provide some clues about how genetic influences on some aspects of smoking behavior may arise.

WHY ARE PEOPLE STILL SMOKING?

The harmful effects of smoking have received widespread publicity beginning in the early 1960s, when government reports documenting many adverse health consequences of smoking began to appear (e.g., National Health & Medical Research Council, 1962; Royal College of Physicians of London, 1962; USPHS, 1964). To better understand why people are still smoking, it is instructive to examine twin-pair resemblance, from surveys in different eras, for current smoking status. By current smoking status, we mean whether or not each twin in a pair is currently smoking cigarettes at the time of a particular survey or whether either twin has never smoked or has quit smoking (grouping together these two latter categories).

The largest data set for examining the genetic involvement in risk of being a current smoker is provided by Medlund et al. (1977), who tabulated data from the New Swedish Twin Register, a register of male and female like-sex twin pairs born throughout Sweden during the period 1926–1967, both of whom were still alive in 1971. Since the register was derived from birth certificates from throughout Sweden, and updated address information was obtained from national records, the register is unusual in its completeness. We present here our own reanalyses of their data. When surveyed in 1973 (6903 female same-sex pairs, 5995 male same-sex pairs), approximately one half of male respondents and 40% of female respondents were current smokers. Smoking in this survey was operationalized as smoking daily, or almost every day, so that occasional "chippers" (Shiffman, 1989) would have been classified as non-smokers. If one twin from a pair was a current smoker, the probability that the co-twin was also a current smoker was greatly increased and, furthermore, the probability was higher if the co-twin was monozygotic (MZ) than if the co-twin was dizygotic (DZ). The so-called probandwise concordance rate, i.e., the probability that the co-twin of a current smoker would also be a current smoker, was 77.3% for MZ male pairs, 66.5% for DZ male pairs, 72.8% for MZ female pairs, and 59.5% for DZ female pairs. In each case, the concordance rate was significantly elevated in MZ compared to DZ pairs, consistent with a genetic influence on risk of being a current smoker.

Raw probandwise concordance rates have no simple interpretation in terms of genetic and environmental parameters, since they are heavily dependent on the prevalence of the phenotype under study (e.g., the proportion of current smokers) in a given sample. (If the fact that one twin is a current smoker does not increase the probability that the co-twin will also be a current smoker, the probandwise concordance rate will simply be equal to the prevalence of current smoking, i.e., the proportion of current smokers.) This is an important issue, since attempts made to interpret concordance rate data (e.g., Hughes, 1986; Carmelli et al., 1992) have led to an understatement of the importance of genetic influences on smoking behavior. However, if we assume that current smoking status is determined by multiple genetic and environmental risk factors that combine additively in their effects, then we may estimate genetic and environmental effects on risk of being a current smoker by fitting a "threshold" model (e.g., Pearson, 1900; Falconer, 1965), which assumes that a continuous normal liability distribution underlies the observed binary distribution of never or ex-smokers vs. current smokers, with those individuals whose "liability" exceeds some critical threshold value becoming smokers. These are the standard assumptions used by geneticists in studying the inheritance of polygenic disorders (i.e., influenced by multiple genetic factors) and also used by psychometricians in the estimation of tetrachoric or polychoric correlations (Olsson, 1979; Olsson, Drasgow, & Dorans, 1982). They have been used extensively, for example, in studies of the genetic influence on alcoholism risk (for further details, see Heath et al., 1994b).

Using these assumptions, we may estimate twin-pair correlations for being a current smoker from the New Swedish Twin Register data, using the numbers of twin pairs who are (1) concordant current smokers; (2) discordant, with one twin a current smoker and the co-twin either an ex-smoker or never having smoked; and (3) concordant current nonsmokers. Strictly, these are correlations for "liability" or risk of being a current smoker, since these are correlations for the underlying liability measure, not the observed binary measure of current vs. ex- or never smoker. Estimated tetrachoric correlations (± standard errors) are: MZ male pairs, 0.78 ± 0.02; DZ male pairs, 0.50 ± 0.02; MZ female pairs; 0.79 ± 0.02; DZ female pairs, 0.52 ± 0.02. The highly significant differences between MZ and DZ tetrachoric correlations for current smoking status confirm the evidence for an important genetic influence in both men and women.

We may proceed one step further by quantifying the contribution to risk of being a current smoker of (1) genetic effects, (2) environmental effects shared equally by twin pairs growing up in the same home (e.g., parent, older sibling, neighborhood, school, or shared peer influences on smoking), and (3) unique environmental effects (sometimes called within-family environmental effects), i.e, those environmental effects that are experienced by only one member of a twin pair. In cross-sectional data, measurement error will be confounded with unique environmental effects, provided that independent assessments have been made of each twin. Once again, estimates of these genetic and environmental effects are obtained by fitting a threshold model (Eaves, Last, Young, & Martin, 1978; Eaves, Eysenck, & Martin, 1989). For current purposes, we ignore the effects of genetic nonadditivity (e.g, genetic dominance or gene–gene multiplicative interactions—epistasis, since these effects would be masked by shared family environmental effects in data on twin pairs reared together (Neale & Cardon, 1992). We make a strong assumption that the relevant environments experienced by MZ twin pairs are no more highly correlated than the environments experienced by DZ twin pairs (we will reexamine this assumption later). Under these assumptions, we find that genetic effects account for 55% of the variance in risk of being a current smoker in women (56% in men), shared environmental effects for 24% of the variance in women (22% in men), and within-family environmental effects for the remaining 21% and 22% of the variance. These estimates were obtained by fitting models to the twin-pair contingency tables for smoking status by the method of maximum likelihood (Eaves et al., 1978). However, since sample sizes are very large, good approximations to the maximum-likelihood estimates can be obtained by (1) doubling the difference between MZ and DZ tetrachoric correlations to estimate the additive genetic variance; (2) subtracting the MZ correlation from twice the DZ correlation to estimate the shared environmental variance; and (3) subtracting the MZ correlation from unity to estimate the within-family environmental variance (Eaves, 1982).

These reanalyses of the New Swedish Twin Register data are consistent with the interpretation that in Sweden in the 1970s, there was a very substantial genetic contribution to risk of being a current smoker. By way of comparison, most studies of the inheritance of personality differences (e.g., Eaves et al., 1989; Loehlin, 1992) have reported estimates of the proportion of the total variance in personality scores attributable to genetic effects (the "heritability" of personality) in the range of 40–50%, i.e., somewhat lower than we observe for current smoking status. Unique environmental effects play a relatively modest role: 80% of the variance in risk of being a current smoker is determined by risk factors (genetic and environmental) shared equally by MZ twin pairs reared in the same household and only 20% by unique environmental effects. This proportion may be contrasted with the much stronger unique environmental effect on personality differences (Plomin & Daniels, 1987). For personality traits, unique environmental effects plus measurement error have typically been found to account for 50–60% of the variance in personality scores (Eaves et al., 1989; Loehlin, 1992), so that even if we allow for greater measurement error in the assessment of personality, the difference in the importance of unique environmental effects is striking. Again contrasting with findings for personality traits—which appear, on the basis of twin, adoption, and separated twin studies, to be almost entirely free of influences of family background (Eaves et al, 1989; Loehlin, 1992)—shared environmental influences on current smoking status are also substantial. Some of this shared environmental influence may be explained by differences in the proportion of individuals who are current smokers as a function of year of birth (for which twin pairs are, of course, perfectly correlated), but this effect is unlikely to be strong, given the modest differences in prevalence of current smoking among different age groups in this sample, particularly in men (Medlund et al., 1977).

How Generalizable Are These Findings?

Let us now examine the extent to which these findings from the 1970s for current smoking status in the New Swedish Twin Register are replicated in different eras and different countries. What can we learn from studies that predate the Medlund et al. (1977) survey? Fisher (1958, 1959) gave the earliest reports of English and German twin data on smoking, in support of his controversial argument that smoking and lung cancer are associated not causally, but through a shared genetic vulnerability. (Fisher's work may be considered an important historical factor in the lack of attention given to the genetic involvement in smoking, since such studies became associated with the view that smoking has no adverse health consequences.) However, the numbers of pairs assessed were very small, and the sample was one of convenience. Raaschou-Nielsen (1960) reported smoking data on a series of 894 MZ and DZ Danish twin pairs, both female and male, born 1870–1910 and identified from birth records. Occasional smokers, smoking less than one cigarette per day, were classified as ex-smokers.

Reanalyzing her data, we find significantly elevated probandwise concordances for current smoking in MZ compared to DZ twin pairs and obtain estimates of genetic and environmental parameters indicating that genetic effects accounted for 78% of the variance in risk of being a current smoker among women, and 72% for men, with the remaining variance accounted for by within-family environmental effects (i.e., there was no significant influence of shared family environment).

Cederlof et al. (1971) reported data for the Old Swedish Twin Register, surveyed initially in 1959–1961 (birth years 1886–1925; 4566 male same-sex pairs and 5754 female same-sex pairs), and for the United States National Academy of Sciences–National Research Council Twin Registry surveyed in 1967–1969 (birth years 1917–1927, identified through World War II military service records linked to birth certificate records; 3875 male same-sex pairs), but present their data in a form in which we cannot recover information about current smoking status. However, Carmelli et al. (1992) reported a reanalysis of the 1967–1969 NAS-NRC Twin Registry data, based on somewhat larger sample sizes (4755 pairs), from which we can derive twin-pair probandwise concordance rates for current smoking status (counting only cigarette smokers as current smokers: 73.8% for MZ male twins, 63.7% for DZ male twins, compared to a prevalence of current smoking in the sample of 52–53%). From these estimates, we obtain a heritability estimate for current smoking of 65%, with shared environmental influences on current smoking status being nonsignificant.

The early large-sample studies thus support the finding of a genetic influence on current smoking status from the New Swedish Twin Register sample, but do not find significant shared environmental influence. Because of the age of the Danish sample, selective attrition of smokers (through death) may have occurred and led to an underestimation of the importance of shared environmental effects (Heath, unpublished simulation). This, however, is a less likely explanation for the similar findings in the somewhat younger United States sample.

More recent twin studies also support an important genetic influence on current smoking status. Kaprio et al. (1978) reported smoking data from a survey in 1975 of a nationwide, birth-certificate-derived twin register in Finland (4936 male same-sex, 5545 female same-sex pairs born 1920–1958 and alive in 1967). Once again, occasional smokers were included with non-smokers. Reanalysis of these data confirms significantly elevated probandwise concordance rates for current smoking in MZ compared to DZ twin pairs, in both genders, and yields estimates of the genetic contribution to risk of being a current smoker of 56% in men and 33% in women, with shared environmental influences accounting for, respectively, 15% and 45% of the variance, and unique environmental effects for 29% and 22%. Thus, the estimate of the importance of genetic factors for Finnish men is comparable to that reported for both men and women for the New Swedish Twin Register. The estimate for Finnish women is much lower, though still substantial and highly significant.

Smoking data from a survey of the Virginia Twin Resister, a birth-certificate-derived register of twin pairs born in the state of Virginia (1915–1967) and surveyed in 1985–1987, have been reported (Heath et al. [1993a]; see also Kendler et al. [1993]: 457 MZ female [MZF], 301 MZ [MZM], 380 DZ female [DZF], 352 DZ male [DZM], and 636 DZ opposite-sex pairs). This twin panel shares with Scandinavian twin registries the advantage of having been systematically ascertained, but is less comprehensive because twin pairs moving out of state were not tracked. Genetic effects account for 74% of the variance in risk of being a current smoker in men and women, with shared environmental effects being non-significant. However, there is a trend for a higher heritability in men (70%) than in women (51%), with shared environmental effects accounting for 25% of the variance in women (but having no significant effect in men) when parameters specific to each sex were estimated.

Data from recent surveys of volunteer twin panels also support an important genetic influence on current smoking status. Data from the Maudsley volunteer twin register in the United Kingdom conducted in the late 1970s (Eaves & Eysenck [1980]; Eaves et al. [1989]: 236 MZF, 80 MZM, 123 DZF, 50 DZM, and 58 DZ opposite-sex pairs) yielded heritability estimates for current smoking status of 63% in men and 38% in women (though it should be noted that these two estimates do not differ significantly because of the very small sample sizes used in that study). Data from the Australian volunteer twin panel, based on a survey of 3808 pairs conducted in 1980–1981 (1233 MZF, 567 MZM, 751 DZF, 352 DZM, and 907 DZ opposite-sex pairs), are reported by Hannah, Hopper, and Mathews (1985) and Heath and Martin (1993). Reanalyzing these data, we again find highly significant genetic effects on current smoking status, accounting for 62% of the variance in liability in males and 48% of the variance in liability in females. Shared environmental influences explain 24% of the variance in females, but do not differ significantly from zero in males. Unique environmental effects explain 28% and 36% of the variance in females and males, respectively.

Overall, we may conclude, on the basis of twin surveys of a variety of Western nations (including the United Kingdom, Australia, and the United States, as well as Finland, Sweden, and Denmark) covering a wide time period (from the late 1950s to the mid-1980s), that twin data are consistent in suggesting an important genetic influence on risk of being a current smoker. Estimates of the heritability of current smoking status are quite variable among studies, ranging from 33% to 78% for women and from 55% to 72% for men, with a trend for lower estimates in women than in men. Some of this variability most probably reflects differences in the operationalization of current smoking status, with some studies excluding occasional smokers and others including such individuals as current smokers. Nonetheless, differences in estimates among studies at the same time period using similar assessment procedures (e.g., comparing the Finnish, Swedish, and NAS-NRC results) suggest that there are important cross-cultural differences in the relative influence of genes and family background on current smoking status.

WHO BECOMES A SMOKER?

The finding of an apparent important genetic influence on the probability that an individual will be a current smoker leaves unanswered very basic questions about how that genetic influence arises. It might be the case that there are genetic influences on risk of becoming a smoker, with an individual's smoking career, once smoking has begun, largely determined by environmental factors. Alternatively, it is possible that whether or not an individual starts to smoke (smoking "initiation") is largely determined by environmental factors—for example, influenced by the behavior of parents, older siblings, and peers (cf. Chassin, Presson, Sherman, Cortz, & Olshavsky, 1984; Mittelmark et al., 1987; Ogawa, Tominaga, Gellert, & Acki, 1988; Swan, Creeser, & Murray, 1990) (see also Chapter Four)—but that after exposure to the effects of nicotine, genetic factors play a more important role in the progression of the smoking habit. Let us consider first the case for an important genetic involvement in risk of becoming a smoker.

In the Swedish survey of Medlund et al. (1977), approximately two thirds of men born 1926–1945 and 60% of men born 1946–1958 have smoked at least 5–10 packs of cigarettes (the definition used to operationalize smoking in that survey) at some stage in their lives. Among women, approximately 40% of those born 1926–1935, but 55% of those born 1936–1958, are current or former smokers. The probandwise concordances for smoking initiation—i.e., the probability that the co-twin of a current or former smoker will also have smoked at some stage—is substantial in both MZ and DZ pairs, but still significantly higher in MZs than in DZs (MZF pairs, 81.7%; DZF pairs, 72.6%; MZM pairs, 88.1%; DZM pairs, 79.2%). Corresponding MZ and DZ tetrachoric correlations are 0.85 $\pm$ 0.01 and 0.64 $\pm$ 0.02 for female pairs and 0.90 $\pm$ 0.01 and 0.64 $\pm$ 0.01 for male pairs, once again confirming the significant genetic contribution to risk of starting to smoke. By model-fitting, we find that genetic factors account for 44% of the variance in risk of becoming a smoker in women vs. 51% in men; shared environmental effects account for, respectively, 42% and 39% of the variance in risk; within-family environmental effects account for the remaining 14% and 10% of the variance.

Comparing findings for smoking initiation in the New Swedish Twin Register sample to those for current smoking status, we find that the genetic influence on smoking initiation is somewhat more modest (albeit still very strong); the shared environmental influence is almost twice as strong as observed for current smoking, and unique environmental effects are even less important. The relative lack of importance of unique environmental effects on smoking initiation parallels findings for use of alcohol (vs. lifetime abstinence) (Heath, Jardine, & Martin, 1989; Heath, Todorov, Madden, Bucholz, & Dinwiddie, 1993b) and for initiation of illicit drug use (Tsuang et al., 1992; Heath et al., 1993b). The strong genetic influence is also consistent with findings for initiation of illicit drug use, in contrast to the relative lack of genetic influence on lifetime abstinence from use of alcohol. From a behavior genetic perspective, however, it is once again the strong shared environmental influence on smoking initiation that is most striking.

How Generalizable Are The Findings for Smoking Initiation?

From the Raaschou-Nielsen data from the late 1950s, combining occasional smokers with never-smokers, we obtain estimates of the genetic contribution to smoking initiation of 79% for women and 84% for men, with no significant shared environmental influences on initiation. For the Old Swedish Twin Register data, we find estimates of the genetic contribution to risk of smoking initiation of 39% for women and 57% for men, with highly significant shared environmental influences accounting for an additional 40% of the variance in women and 21% of the variance in men. For the male NAS–NRC World War II twin panel, from the data reported by Carmelli et al. (1992), we obtain a heritability estimate of 59%, with shared environmental effects accounting for an additional 21% of the variance. For the Finnish twin register data, we obtain heritability estimates of 37% for women and 50% for men, with shared environmental effects accounting for, respectively, 50% and 33% of the variance in risk of becoming a smoker. For the London twin sample, neither the hypothesis of no genetic influence nor the hypothesis of no shared environmental influence could be rejected, because of small sample sizes, but under a model allowing for both genetic and shared environmental effects, a heritability estimate of 38% was obtained for both men and women with shared environmental effects accounting for an additional 29% of the variance. For the Australian twin panel, heritability estimates of 28% for men and 77% for women were reported by Heath et al. (1993a) with shared environmental effects accounting for 43% and for a nonsignificant 4% of the variance, respectively. For the Virginia twin panel, heritability estimates for men and women of 84% were obtained, once again with no significant shared environmental influence on risk of becoming a smoker (Heath et al., 1993a).

Two adolescent twin studies have also obtained data on smoking initiation. In a survey of adolescent and young adult Dutch twins, Boomsma, Koopmans, van Doornen, and Orlebeke (1994) reported a heritability of smoking initiation of 31%, with shared environmental factors accounting for an additional 59%. Hopper, White, Macaskill, Hill, and Clifford (1993) also reported higher MZ than DZ concordances in an analysis of data from adolescent Australian twins, but reported nonparametric analyses of their data from which genetic and environmental parameter estimates cannot be derived. Regarding both these data sets, it should be noted that since they are derived from adolescents, the possibility that results will change as the samples grow older cannot be excluded.

Findings from these surveys of adult twins, using samples ranging in date of birth from 1870 to 1967, are very consistent in indicating a highly significant genetic contribution to risk of smoking initiation. In this respect, they confirm the findings from the New Swedish Twin Register. In other respects, however, results are remarkably variable. Heritability estimates for women range from 37% to 84%, those for men from 28% to 84%. Most studies find either no substantial gender difference in heritability or else a higher heritability in men than in women, but in the Australian sample this trend is reversed. Studies in Finland, Sweden, the United States (NAS–NRC twin panel only), and Australia (men only) indicate a strong shared environmental influence on risk of smoking initiation, while results from Denmark and the United States (Virginia twin panel) do not find significant shared environmental influences. In contrast to the marked cross-cultural consistency of findings for mode of inheritance of personality traits (e.g., Eaves et al., 1989; Loehlin, 1992), smoking initiation appears to be a far more "plastic" trait (cf. Cavalli-Sforza & Feldman, 1981) with a mode of inheritance that is quite variable across cultures. Unexpectedly, however, the one paper that tested for changes in mode of inheritance within cultures over time (comparing data from the United States and Australia for birth years spanning the 20th century through 1967 [Heath et al., 1993a]) failed to find significant heterogeneity of genetic or environmental influences as a function of year of birth.

Can We Exclude Nongenetic Explanations?

Only limited separated twin and adoption data are available for smoking variables, and such data are typically based on small sample sizes. For example, Jules-Nielsen (cited in Raaschou-Nielsen, 1960) reported smoking data on 12 separated MZ twin pairs. Eaves and Eysenck (1980) reported data on adoptive siblings and adoptive parent–child pairs, but with no information on the biological parents of adopted-away children. In the absence of large-sample data from adoption studies or separated-twin studies, the evidence for a genetic influence on smoking initiation rests chiefly on the comparison of twin-pair concordances in MZ vs. DZ twin pairs reared together. There are a number of environmental experiences for which MZ twin pairs are more likely to be concordant than DZ twin pairs: they are more likely to share the same friends, more likely to be dressed alike, more likely to maintain close social contact with one another as adults, and so on (e.g., Loehlin & Nichols, 1976; Madden et al., 1993a). Could it be that the higher MZ than DZ concordances that we observe for smoking initiation are merely a function of greater sharing of environmental risk factors? Given the widespread belief in the importance of peer influences on smoking initiation, the greater sharing of same-sex peer friends by MZ than by DZ same-sex pairs, in particular, is cause for concern.

To address this issue, Madden et al. (1993a) examined twin-pair tetrachoric correlations separately for those pairs who reported usually or always sharing the same friends when growing up and those pairs who reported sometimes or always having different friends. As would be predicted under the peer-influence hypothesis, twin correlations were much higher in those pairs who usually shared the same friends than in those who did not. However, in both groups, MZ correlations were significantly elevated compared to same-sex DZ correlations, and estimates of the genetic contribution to smoking initiation remained substantial, indicating that the apparent genetic influence could not be explained away by peer effects.

While the finding of higher twin-pair correlations for those who shared the same peers is consistent with an important peer influence, alternative explanations could not be excluded—for example, that there is a selective friendship effect, with smokers being more likely to associate with other smokers, but no direct causal environmental influence of peers' smoking. For other reported aspects of environmental experience, significant differences in twin-pair tetrachoric correlations between similar-environment and dissimilar-environment groups were found only for sharing the same room with co-twin as children.

One important potential environmental influence that must be considered is the behavior of the co-twin. Is it possible that smoking by one twin has a direct environmental influence on the probability that the co-twin will also become a smoker? In analyses of data from Australian and United States samples (Heath et al., 1993a), we did indeed obtain results consistent with an important twin reciprocal environmental influence. However, estimates of the genetic contribution to probability of smoking initiation remained substantial (46–77%), and it was the magnitude of the estimate of the shared environmental influence that was decreased. Indeed, once allowance was made for these reciprocal environmental influences of one twin on the other, it was not possible to reject the hypothesis that there were no other significant shared environmental influences on probability of smoking initiation (Heath et al., 1993a).

How Do We Explain the Genetic Influence?

Further confirmation of the genetic influence on smoking initiation is provided when we look for mechanisms by which such an influence might arise. The involvement of genetic factors in the inheritance of personality differences, and lack of importance of shared environmental experiences, is well established. Early work by Eaves and Eysenck (1980) failed to find strong relationships between smoking and personality differences, assessed using the Eysenckian personality dimensions (Extraversion, Neuroticism, Toughmindedness, and Social Conformity). Data from the Australian twin panel, however, confirm a substantial genetic correlation between the personality trait Novelty Seeking, assessed using the Tridimensional Personality Questionnaire (Cloninger, Przybeck, & Svrakic, 1991; Heath, Cloninger, & Martin, 1994a), and risk of becoming a smoker (Madden et al., 1993b,c), suggesting that impulsive or risk-taking personality traits may be important mediators of the genetic influence on probability of becoming a smoker. In data on a young panel of Australian twins aged 18–25 when surveyed in 1989–1990, smokers were higher in Novelty Seeking than never-smokers, but also more Extraverted and more Neurotic (Heath & Madden, 1993). Additionally, in women, at least, a significant residual genetic correlation was found between Neuroticism and risk of smoking initiation, even when differences in Novelty Seeking were controlled for. Genetically determined personality differences nonetheless accounted for only 37% of the total genetic variance in risk of smoking initiation in women and for 55% in men, leaving a substantial proportion of the genetic variance unexplained.

WHY DO SMOKERS KEEP SMOKING?

Finally, let us turn to the question of whether there are genetic influences on progression of the smoking habit among smokers. From the perspective of health consequences and the economic costs of smoking, the most important phenotype here is that of smoking "persistence" (Eaves & Eysenck, 1980), i.e., whether an individual who has started to smoke regularly keeps smoking or quits successfully. All other constructs (e.g., "dependence," amount smoked) are of interest primarily insofar as they predict this outcome. Once again, let us turn back to the New Swedish Twin Register data to examine whether there are genetic influences on persistence.

The Swedish data have some drawbacks, since they are cross-sectional, and we have no information about how many years the twins have smoked. Nonetheless, since onset of smoking typically occurs in adolescence or early adulthood, in the older cohorts, smokers will have been smoking for many years. Among male respondents to the 1973 Swedish survey, the proportion of smokers still smoking regularly (operationalized as daily or nearly every day) ranged from 74% in the oldest cohort to 83% in the youngest. Among female smokers, corresponding proportions ranged from 73% to 79%. Given these substantial prevalences for persistence in smoking among lifetime smokers, large sample sizes are needed to detect a significant genetic influence on smoking persistence. Nonetheless, probandwise concordance rates for persistence (considering only those pairs in which both twins had become cigarette smokers) were significantly elevated above the population prevalence and significantly elevated in MZ pairs compared to DZ pairs. Probandwise concordance rates were 86% for MZF pairs, 81% for DZF pairs, 86% for MZM pairs, and 83% for DZM pairs. Corresponding tetrachoric correlations ($\pm$ standard errors) were, for female same-sex pairs, 0.59 ± 0.04 for MZ and 0.31 ± 0.04 for DZ pairs, and, for male same-sex pairs, 0.49 ± 0.05 for MZ and 0.31 ± 0.04 for DZ pairs. Model-fitting (using only data from pairs in which both twins had smoked at some stage in their lives) yielded heritability estimates of 59% for women and 52% for men, with no significant shared environmental influences, and the remaining 41% and 48% of the variance being explained by within-family environmental effects.

Analyses of smoking persistence using data only from pairs in which both twins have become smokers must be interpreted with some caution, since they implicitly assume that genetic and environmental determinants of smoking persistence are statistically independent of the genetic and environmental determinants of smoking initiation (Heath, 1990; Heath & Martin, 1993). If in fact there is substantial overlap between genetic influences on initiation and genetic influences on persistence (as we would predict if impulsive risk-taking traits were influencing both these measures), then using only data from pairs concordant for ever having smoked would lead to biases to estimates of genetic and environmental parameters (Heath, 1990). However, we may examine this critical assumption by analyzing data from the 3×3 contingency tables computed for each twin group, using the classification never smoked/ex-smoker/current smoker; and may compare the fit of models that assume either (1) a multiple threshold model, in which successful quitters are intermediate in liability between persistent smokers and never smokers ("one-process" model); or (2) independent determinants of initiation and persistence, with genetic and environmental influences on persistence observable only in those individuals who become smokers ("two-process" model); or (3) a model combining elements of both (1) and (2) in which some ex-smokers are "early quitters" who are low in liability on the initiation dimension, and others are successful quitters who have become regular smokers, but are low in liability on the "persistence" dimension.

This combined model is represented diagrammatically as a probability tree in Fig. 1. Comparing the goodness of fit of these three models, we find that the first two models are rejected at a very high significance level ($p < 0.01$ in both women and men), while the combined model gives a substantially better (albeit still somewhat marginal) fit to the data ($p = 0.01$). Estimates of the genetic contribution to smoking persistence are actually *increased* over those obtained when we simply ignore data from pairs in which one twin had never smoked.

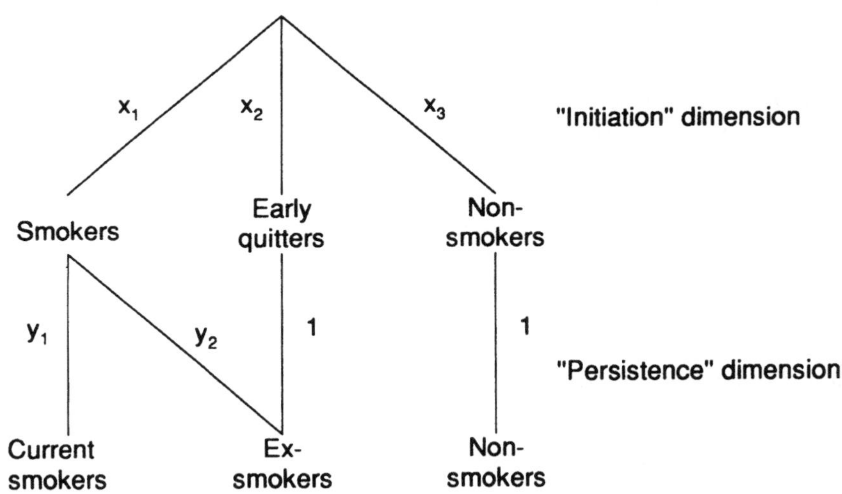

FIGURE 1. Combined two-process model for smoking initiation and smoking persistence.

Under the best-fitting combined model, genetic factors account for 58% of the variance in smoking persistence among Swedish male smokers and 70% of the variance in Swedish female smokers, with the remaining variance accounted for by within-family environmental effects plus measurement error.

How Generalizable are the Findings for Smoking Persistence?

Sample sizes in the Raaschou-Nielsen study were relatively small. For males, we obtain a heritability estimate for smoking persistence of 47%, but this does not differ significantly from zero. For females, numbers of pairs in which both twins were current or former smokers and at least one had quit smoking are too small to permit meaningful analysis. Data from the Old Swedish Twin Register and the NAS-NRC sample reported by Cederlof et al. (1971) are not presented in a form that is informative about genetic influence on smoking persistence. Although the recent paper by Carmelli et al. (1992) reanalyzing the NAS-NRC sample aimed to be informative about "genetic effects on . . . the ability to quit smoking," neither analyses nor summary data presented in that paper allow the discrimination of genetic effects on persistence from genetic influences on smoking initiation. (For this purpose, we need data on concordance rates for persistence among pairs concordant for ever having smoked, data which cannot be reconstructed from their paper.)

From reanalyses of the Finnish twin data, we obtained heritability estimates for smoking persistence of 68% for men and 71% for women, with no significant shared environmental influence in either gender. In their analysis of the 8-year follow-up of the Australian NH&MRC twin panel, Madden et al. (1993a) reported a heritability of smoking persistence of 62% for both men and women under a two-process model, that estimate increasing to 74% under the better-fitting combined model. No evidence for shared environmental effects was found in either case. Data from the Virginia twin panel are consistent with a heritability for smoking persistence under a two-process model of 58% in both men and women, but once again this estimate does not differ significantly from zero because of small sample sizes. From the large-sample studies, at least, we thus find evidence consistent with a strong genetic influence on smoking persistence, but no significant shared environmental influence.

How Do We Explain the Genetic Influence on Smoking Persistence?

In contrast to the associations observed between smoking initiation and personality, we have not found strong correlations between smoking persistence and personality, once nonsmokers are excluded from the analysis (Heath & Madden, 1993). In data from the young 18 to 25-year-old cohort in the 1989 survey of the Australian twin panel, only a very modest genetic correlation was observed between smoking persistence and Novelty Seeking, with genetic influences on Novelty Seeking accounting for less than 5% of the total genetic variance in smoking persistence in either gender (Heath & Madden, 1993).

It is possible that once an individual becomes a regular smoker, genetically determined differences in reactions to nicotine (or other pharmacologically active agents contained in cigarettes) take over control of the course of the smoking habit. Several analyses have tested for a genetic influence on number of packs per day smoked, although the data available have important shortcomings. In some reports, nonsmokers are included as a zero point on this scale (e.g., Kaprio et al., 1982), thereby confounding genetic influences on initiation with genetic influences on amount smoked, two processes that appear to be under relatively independent genetic control (Meyer, Heath, & Eaves, 1992). Some reports give only the numbers of concordant pairs at each level of daily cigarette consumption, rather than the full two-way contingency table cross-classifying daily cigarette consumption levels of the twins in the pairs (e.g., Medlund et al., 1977; Carmelli et al, 1992). Studies also differ in whether they report information about quantity smoked only for current smokers (which will lead to biased estimates of genetic and environmental parameters if genetically determined differences in daily consumption affect probability of successful smoking cessation) or for both current and former smokers (in which case the possibility of inaccurate reporting by ex-smokers may be a problem).

Data from the New Swedish Twin Register are not presented in a form that would permit separate analysis of the genetic contribution to level of cigarette consumption. From the Raaschou-Nielsen data, under the assumption of approximately independent inheritance of smoking initiation, and of quantity smoked by those who become smokers, we may compute for those twin pairs who are both current regular smokers contingency tables for the binary "quantity" variable of heavy smoking, operationalized in that survey as smoking 15 or more cigarettes per day. The numbers of female same-sex pairs in which both twins are regular or heavy smokers are too few to be usable, but for quantity smoked in males, we obtain a heritability estimate of 61%, with no significant shared environmental influence. The paper by Carmelli et al. (1992) again does not allow us to separate genetic effects on quantity smoked from genetic effects on smoking initiation, since it is not possible to separately compute numbers of pairs discordant for quantity smoked where both are current smokers and numbers of pairs discordant because only one twin is a current smoker.

Kaprio et al. (1978) reported twin-pair intraclass correlations for the number of "pack years" smoked (i.e., packs per day × years smoked). This measure is highly confounded with age (which will mimic shared environmental effects in twin data), since younger smokers will on average have smoked for fewer years, but nonetheless for current smokers exhibits substantial genetic influence (heritabilities of 55% for women and 38% for men). For the Virginia twin register, Meyer and colleagues (personal communication) found a heritability estimate for quantity smoked of 69%, in an analysis under a two-process model allowing for independent genetic and environmental determination of smoking initiation and level of smoking by those who become smokers. In the only data that have been reported for the Australian twin panel, based on the 1981 twin cohort resurveyed in 1989, more modest estimates of the heritability of quantity smoked were obtained (18% in men, 37% in women) (Heath & Madden, 1993), but with a substantial genetic correlation between genetic influences on quantity smoked and genetic influences on smoking persistence.

DISCUSSION

The data that we have reviewed seem to confirm an important genetic influence on all aspects of smoking behavior, including risk of becoming a smoker, amount smoked, and the probability of persisting in the smoking habit and becoming a long-term smoker. For smoking initiation, at least some of the observed genetic influence may be mediated through personality variables related to risk-taking or impulsiveness, although it seems likely that there is substantial residual genetic variation that cannot be explained by the inheritance of personality differences. It is possible that further genetic variance will be accounted for by genetic influences on sociodemographic variables that predict risk of smoking initiation (e.g., educational attainment [Vogler & Fulker, 1983]). Furthermore, the magnitude of the genetic influence on smoking initiation may have been overestimated in analyses of twin data that failed to allow for the greater similarity of early environmental influences (particularly shared friends) in MZ compared to DZ twin pairs. It remains to be determined how much residual genetic variance needs to be accounted for, once personality and sociodemographic effects on probability of smoking initiation are allowed for.

In contrast, when we consider the genetic contribution to probability of smoking persistence, and to number of packs per day smoked, we find evidence for a substantial genetic influence that cannot be explained by personality or sociodemographic influences. Perhaps it is the case that once regular smoking has begun, genetic differences in response to nicotine (or other substances contained in tobacco) are largely responsible for determining progression of the smoking habit. Independent evidence for genetic control of response to nicotine is provided by studies of nonhuman species. Collins and his collaborators (reviewed in Collins & Marks, 1989) have systematically examined the genetics of first-dose sensitivity and the capacity to develop tolerance to nicotine using behavioral and physiological responses to administrations of nicotine among inbred mouse lines. Nicotine has been observed to affect many systems, and response to dosing with nicotine in mice has been measured using (1) a battery of tests devised to assess changes in several parameters including motor activity, body temperature, rate of respiration, acoustic startle response, and latency to seizure; and (2) assessments of nicotinic receptor binding using two ligands, L-[^{3}H]nicotine and β[^{125}I]bungarotoxin. Strain differences in response to an acute dose of nicotine in naïve animals have been a consistent finding, suggesting that initial sensitivity to nicotine is regulated to some extent by genetic factors. Furthermore, strains demonstrating greater initial sensitivity have been found to possess a greater number of nicotinic brain receptors compared with strains observed to be more resistant.

Genetic influence on the development of tolerance has been examined using strains of mice that demonstrate broad differences in sensitivity to nicotine (reviewed in Collins & Marks, 1989; Marks, Stitzek, & Collins, 1986). Animals continuously infused with either a saline solution or one of several increasing doses of nicotine over a period of days have been tested for tolerance using acute doses of nicotine 2 hours after stopping infusion (to allow for elimination). Levels of tolerance developed in mice of strains that manifest greater initial sensitivity have demonstrated dose dependence, while changes in tolerance for strains less reactive to first-dose nicotine have been found to occur at higher chronic doses (indicating a higher threshold for drug response) and to be markedly less by comparison. Strain differences in response have not been consistently associated with differences in receptor activity; with chronic infusion, increases in receptor binding have been found to be similarly dependent on chronic dose across mouse strains. The extent to which differences in tolerance in mice can be explained by differences in central nervous system performance remains unclear.

We noted at the beginning of this chapter the relative paucity of human studies directly targeted at understanding the genetic contribution to smoking behavior, contrasted with the many studies that have tried to establish a genetic contribution to alcoholism risk (see Chapter 2). Evidence for a genetic influence on alcoholism risk has stimulated studies of high-risk individuals designed to elucidate mechanisms by which such an influence might arise, ranging from investigation of the response to a challenge dose of alcohol of individuals at elevated alcoholism risk and controls (see, for example, the meta-analysis of such studies reported by Pollock, 1992) through biochemical studies (e.g., Tabakoff, Whelan, & Hoffman, 1990) to evoke potential studies (e.g., Begleiter, Porjesz, Bihari, & Kissin, 1984). It is arguable that the case for a substantial genetic involvement in smoking behavior is much stronger than the case for such an involvement in alcoholism. Although most studies report only point estimates for the heritability of alcoholism, in Heath et al. (1994b), 95% confidence intervals were reported, which for many widely cited studies are surprisingly broad. In Fig. 2, we give point estimates and confidence intervals for the genetic involvement in smoking behaviors derived from the data of Medlund et al. (1977), Kaprio et al. (1978), and Heath and Martin (1993), i.e., the three largest data sets that present analyzable data on current smoking status, smoking initiation, and smoking persistence. While confidence intervals for smoking persistence in particular are somewhat broad (because only data on twin pairs concordant for lifetime smoking are used), the consistency of the evidence for an important genetic influence is striking.

Given the strength of the evidence for a genetic involvement in smoking

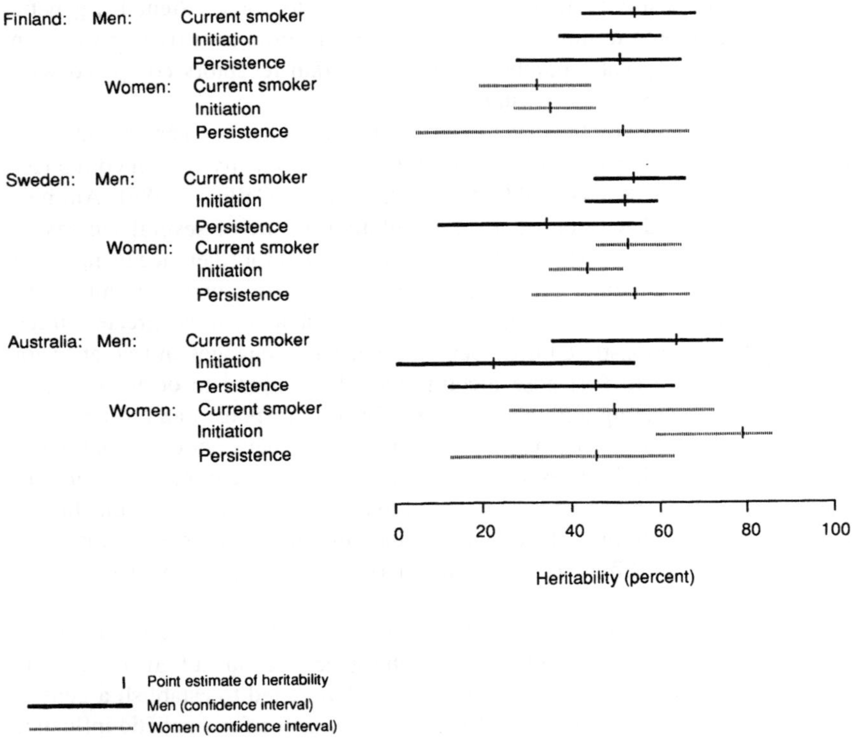

FIGURE 2. Heritability estimates and 95% confidence intervals for smoking variables in the Swedish, Finnish, and Australian twin studies (see the text for further details).

persistence, and the crucial importance of this outcome from the perspective of health risks and economic costs, it is natural to wonder how further insight can be obtained into the mechanisms by which these genetic influences arise. Recent refinements of methods of nicotine administration, to allow more consistency in dosing and better approximation of the rapid peak in concentration and decline of nicotine in blood levels observed after smoking a cigarette, offer the prospect of studying the genetic control of nicotine sensitivity, i.e., innate tolerance in naïve human subjects. Pomerleau, for example, has presented data that contrast response to nicotine in smokers and never-smokers (Pomerleau, Collins, Shiffman, & Pomerleau, 1993a; Pomerleau, Hariharan, Pomerleau, Cameron, & Guthrie, 1993b). He reported finding nicotine accumulation in the body, as measured by blood plasma levels following an acute dose of nicotine, to be significantly less for smokers than for nonsmokers (ratio of mean ± SEM of peak plasma nicotine level in nanograms per milliliter to dose of nicotine administered milligrams), suggesting that smokers may have greater pharmacokinetic tolerance, i.e., be more efficient at the distribution and elimination of nicotine. Direct comparison of smokers and nonsmokers cannot tell us, of course, the extent to which such differences in tolerance are acquired vs. innate. However, given the evidence for relatively independent genetic control of smoking initiation vs. smoking persistence (Heath & Martin, 1993; Madden et al., 1993a), it should be possible to examine the contribution to risk of smoking persistence of genetically determined differences in pharmacokinetic tolerance to nicotine using twin pairs discordant for ever having smoked. Specifically, response to nicotine may be examined in never-smoked MZ and DZ co-twins of persistent smokers vs. successful quitters. The hypothesis of genetically determined differences in tolerance that contribute to risk of smoking persistence predicts significant differences between co-twins of persistent smokers vs. successful quitters, which are greater in magnitude in MZ than in DZ co-twins (i.e., greater regression to the mean in the DZ cotwins).

Pomerleau et al. (1993a,b) further suggested that greater initial sensitivity to nicotine is associated with increased capacity to develop tolerance, leading to increased risk of smoking persistence. Physiological reactivity in male never-smokers to acute nicotine exposure was compared with response in age-matched light and heavy smokers who had been nicotine-deprived overnight, and more pronounced physiological reactions were observed in the more dependent smokers. Again, use of smoking-discordant twins offers the possibility of examining genetic control of innate responsiveness to nicotine and its relationship to smoking persistence and the development of tolerance produced after repeated dosing. Although preliminary studies on nicotine response, such as these completed by Pomerleau and colleagues, suggest that differences in first-dose sensitivity to nicotine may better predict capacity to develop tolerance and ultimately, once a smoker, the likelihood of persistent smoking.

A further important area remaining to be addressed is that of the specificity of genetic influences on smoking persistence or their communality with genetic influences on dependence on alcohol or other substances. Joint analyses of alcohol consumption patterns and smoking initiation by Swan et al. (1990) have been extended by analyses including a measure of alcohol-related problems (Madden et al., 1993b,c). The results suggest important associations in genetic risk among these measures of substance use, raising questions concerning the relationships among smoking initiation, liability to persistent smoking, and risks for dependence on alcohol with other types of drugs.

ACKNOWLEDGMENTS. The research reported in this chapter was supported by NIDA postdoctoral training grant DA07261 (PAFM), by NIH grants DA05588, AA07535, and AA07728, and by grants from the Australian National Health and Medical Research Council. Some of the conclusions of this chapter were reported originally at the sixth national ASAM conference on Nicotine Dependence, November 11–14, 1993.

REFERENCES

Begleiter, H., Porjesz, B., Bihari, B., & Kissin, B. (1984). Event-related brain potentials in boys at risk for alcoholism. *Science, 225*, 1493–1496.

Boomsma, D. I., Koopmans, J. R., van Doornen, L. J. P., & Orlebeke, J. F. (1994). Genetic and social influences on starting to smoke: A study of Dutch adolescent twins and their parents. *Addiction, 89,* 219–226.

Carmelli, D., Swan, G. E., Robinette, D., & Fabsitz, R. (1992). Genetic influence on smoking—a study of male twins. *New England Journal of Medicine, 327,* 881–883.

Cavalli-Sforza, L. L., & Feldman, M. (1981). *Cultural transmission and evolution: A quantitative approach.* Princeton, NJ: Princeton University Press.

Cederlof, R., Epstein, F. H., Friberg, L. T., Hrubec, Z., & Radford, E. P. (1971). Twin registries in the study of chronic diseases (with particular reference to the relation of smoking to cardiovascular and pulmonary diseases). *Acta Media Scandinavica Supplementum, 523,* 1–40.

Chassin, L., Presson, C., Sherman, S. J., Cortz, E., & Olshavsky, R. W. (1984). Predicting the onset of cigarette smoking in adolescents: A longitudinal study. *Journal of Applied Social Psychology, 14,* 224–243.

Cloninger, C. R., Przybeck, T. R., & Svrakic, D. M. (1991). The tridimensional personality questionnaire: U.S. normative data. *Psychological Reports, 69,* 1047–1057.

Collins, A. C., & Marks, M. J. (1989). Chronic nicotine exposure and brain nicotinic receptors—influence of genetic factors. *Progress in Brain Research, 79,* 137–146.

Eaves, L. J. (1982). The utility of twins. In V. E. Anderson, W. A. Hauser, J. K. Penry, & C. F. Sing (Eds.), *Genetic basis of the epilepsies* (pp. 249–276). New York: Raven Press.

Eaves, L. J., & Eysenck, H. J. (1980). The genetics of smoking. In H. J. Eysenck (Ed.), *The causes and effects of smoking* (pp. 140–314). London: Maurice Temple Smith.

Eaves, L. J., Eysenck, H. J., & Martin, N. G. (1989). *Genes, culture and personality: An empirical approach.* London: Academic Press.

Eaves, L. J., Last, K., Young, P. A., & Martin, N. G. (1978). Model-fitting approaches to the analysis of human behavior. *Heredity, 41,* 249–320.

Falconer, D. S. (1965). The inheritance of liability to certain diseases estimated from the incidence among relatives. *Annals of Human Genetics, 29,* 51–76.

Fisher, R. A. (1958). Cancer and smoking. *Nature, 182,* 596.

Fisher, R. A. (1959). *Smoking, the cancer controversy: Some attempts to assess the evidence.* Edinburgh: Oliver & Boyd.

Hannah, M. C., Hopper, J. L., & Mathews, J. D. (1985). Twin concordance for a binary trait. II. Nested analysis of ever-smoking and ex-smoking traits and unnested analysis of a "committed-smoking" trait. *American Journal of Human Genetics, 37,* 153–165.

Heath, A. C. (1990). Persist or quit? Testing for a genetic contribution to smoking persistence. *Acta Geneticae Medicae et Gemellologiae, 39,* 447–458.

Heath, A. C., Cates, R. C., Martin, N. G., Meyer, J., Hewitt, J. K., Neale, M. C., & Eaves, L. J. (1993a). Genetic contribution to risk of smoking initiation: Comparisons across birth cohorts and across cultures. *Journal of Substance Abuse, 5,* 221–246.

Heath, A. C., Cloninger, C. R., & Martin, N. G. (1994a). Testing a model for the genetic structure of personality: A comparison of the personality systems of Cloninger and Eysenck. *Journal of Personality and Social Psychology, 66,* 762–775.

Heath, A. C., Jardine, R., & Martin, N. G. (1989). Interactive effects of genotype and social environment on alcohol consumption in female twins. *Journal of Studies on Alcohol, 50,* 38–48.

Heath, A. C., & Madden, P. A. F. (1993). Personality correlates of smoking behavior: A genetic perspective. Paper presented at the American Society of Addiction Medicine Nicotine Dependence Conference, Atlanta, November 11–14.

Heath, A. C., & Martin, N. G. (1993). Genetic models for the natural history of smoking: Evidence for a genetic influence on smoking persistence. *Addictive Behaviors, 18,* 19–34.

Heath, A. C., Slutske, W. S., & Madden, P. A. F. (in press). Gender differences in the genetic contribution to alcoholism risk and drinking patterns. In R. W. Wilsnack & S. C. Wilsnack (Eds.), *Gender and alcohol.* Rutgers, NJ: Rutgers University Press.

Heath, A. C., Todorov, A. A., Madden, P. A. F., Bucholz, K. K., & Dinwiddie, S. H. (1993b).

Modelling the role of genetic factors in the natural history of substance use disorders. Paper presented at the World Congress on Psychiatric Genetics, New Orleans.

Hopper, J. L., White, V. M., Macaskill, G. T., Hill, D. J., & Clifford, C. A. (1992). Alcohol use, smoking habit and the Junior Eysenck Personality Questionnaire in adolescent Australian twins. *Acta Geneticae Medicae et Gemellologiae, 41,* 311–324.

Hrubec, Z., & Neel, J. V. (1978). The National Academy of Sciences—National Research Council Twin Registry: Ten years of operation. In W. E. Nance (Ed.), *Twin research: Part B. Biology and epidemiology* (pp. 153–172). New York: Alan R. Liss.

Hughes, J. R. (1986). Genetics of smoking: A brief review. *Behavior Therapy, 17,* 335–345.

Kaprio, J., Hammar, N., Koskenvuo, M., Floderus-Myrhed, B., Langinvainio, H., & Sarna, S. (1982). Cigarette smoking and alcohol use in Finland and Sweden: A cross-national twin study. *International Journal of Epidemiology, 11,* 378–386.

Kaprio, J., Sarna, S., Koskenvuo, M., & Rantasalo, I. (1978). *The Finnish Twin Registry: Baseline characteristics. Section II.* Helsinki: University of Helsinki Press.

Kendler, K. S., Neale, M. C., MacLean, C. J., Heath, A. C., Eaves, L. J., & Kessler, R. C. (1993). Smoking and major depression: A causal analysis. *Archives of General Psychiatry, 50,* 36–43.

Loehlin, J. C. (1992). *Genes and environment in personality development: Individual differences and developmental series,* Vol. 2. Newbury Park, CA: Sage Publications.

Loehlin, J. C., & Nichols, R. C. (1976). *Heredity, environment, and personality: A study of 850 sets of twins.* Austin: University of Texas Press.

Madden, P. A. F., Heath, A. C., Bucholz, K. K., Dinwiddie, S. H., Dunne, M. P., & Martin, N. G. (1993a). Genetics and smoking. Paper presented at the 55th annual meeting of the College on Problems of Drug Dependence, Toronto, June 12–17.

Madden, P. A. F., Heath, A. C., Bucholz, K. K., Dinwiddie, S. H., Dunne, M. P., & Martin, N. G. (1993b). Novelty seeking and the genetic determinants of smoking initiation and problems related to alcohol use in female twins. Paper presented at the 23rd annual meeting of the Behavior Genetics Association, Sydney, July 13–16.

Madden, P. A. F., Heath, A. C., Bucholz, K. K., Dinwiddie, S. H., Dunne, M. P., & Martin, N. G. (1993c). The genetic relationship between problems related to alcohol use, smoking initiation and personality. Paper presented at the meeting of the Research Society on Alcoholism, San Antonio, June 19–24.

Marks, M. J., Stitzek, J. A., & Collins, A. C. (1986). Dose–response analysis of nicotine tolerance and receptor changes in two inbred mouse strains. *Journal of Pharmacology and Experimental Therapeutics, 239,* 358–364.

Medlund, P., Cederlof, R., Floderus-Myrhed, B., Friberg, L., & Sorensen, S. (1977). A new Swedish twin registry. *Acta Medica Scandinavica Supplementum, 600,* 1–11.

Meyer, J. M., Heath, A. C., & Eaves, L. J. (1992). Using multidimensional scaling on data from pairs of relatives to explore the dimensionality of categorical multifactorial traits. *Genetic Epidemiology, 9,* 87–107.

Mittelmark, M. B., Murray, D. M., Luepker, R. V., Pechacek, T. F., Pirie, P. L., & Pallonen, U. E. (1987). Predicting experimentation with cigarettes: The childhood antecedents of smoking study (CASS). *American Journal of Public Health, 77,* 206–208.

National Health & Medical Research Council (1962). *Smoking and lung cancer: Report of the 53rd Session of the NHMRC.* Canberra: Commonwealth Government Printer.

Neale, M. C., & Cardon, L. R. (1992). *Methodology for genetic studies of twins and families: NATO ASI Series.* Dordrecht, The Netherlands: Kluwer Academic Publishers.

Ogawa, H., Tominaga, S., Gellert, G., & Acki, K. (1988). Smoking among junior high school students in Nagoya, Japan. *International Journal of Epidemiology, 17,* 814–820.

Olsson, U. (1979). Maximum-likelihood estimation of the polychoric coefficient. *Psychometrika, 44,* 443–460.

Olsson, U., Drasgow, F., & Dorans, N. J. (1982). The polyserial correlation coefficient. *Psychometrika, 47,* 337–347.

Pearson, K. (1900). Mathematical contribution to the theory of evolution. VII. On the correlation of character not quantitatively measurable. *Philosophical Transactions of the Royal Society of London, Series, A, 195,* 1–47.

Peto, R., Lopez, A. D., Boreham, J., Thun, M., & Heath, C. (1992). Mortality from tobacco in developed countries: Indirect estimation from national vital statistics. *Lancet, 339,* 1268–1278.

Pierce, J. P., Fiore, M. C., Novotny, T. E., Hatziandreu, E. J., & Davis, R. M. (1989a). Trends in cigarette smoking in the United States: Educational differences are increasing. *Journal of the American Medical Association, 261,* 56–60.

Pierce, J. P., Fiore, M. C., Novotny, T. E., Hatziandreu, E. J., & Davis, R. M. (1989b). Trends in cigarette smoking in the United States: Projections to the year 2000. *Journal of the American Medical Association, 261,* 61–65.

Plomin, R., & Daniels, D. (1987). Why are children in the same family so different from one another? *Behavioral Brain Sciences, 10,* 1–60.

Pollock, V. E. (1992). Meta-analysis of subjective sensitivity to alcohol in sons of alcoholics. *American Journal of Psychiatry, 149,* 1534–1538.

Pomerleau, O. F., Collins, A. C., Shiffman, S., & Pomerleau, C. S. (1993a). Why some people smoke and others do not: New perspectives. *Journal of Consulting and Clinical Psychology, 61,* 723–731.

Pomerleau, O. F., Hariharan, M., Pomerleau, C. S., Cameron, O. G., & Guthrie, S. K. (1993b). Differences between smokers and never-smokers in sensitivity to nicotine: A preliminary report. *Addiction, 88,* 113–118.

Raaschou-Nielsen, E. (1960). Smoking habits in twins. *Danish Medical Bulletin, 7,* 82–88.

Royal College of Physicians of London (1962). *Smoking and health.* New York: Pitman Publishing.

Shiffman, S. (1989). Tobacco "chippers": Individual differences in tobacco dependence. *Psychopharmacology, 97,* 539–547.

Swan, A. V., Creeser, R., & Murray, M. (1990). When and why children first start to smoke. *International Journal of Epidemiology, 19,* 323–330.

Tabakoff, B., Whelan, J. P., & Hoffman, P. L. (1990). Two biological markers of alcoholism. In C. R. Cloninger & H. Begleiter (Eds.), *Banbury report 33: Genetics and biology of alcoholism* (pp. 195–202). Plainview, NY: Cold Spring Harbor Laboratory Press.

Tsuang, M. T., Lyons, M. J., Eisen, S. A., True, W. T., Goldberg, J., & Henderson, W. (1992). A twin study of drug exposure and initiation of use. *Behavior Genetics, 22,* 756 (abstract).

USDHEW (U.S. Department of Health, Education & Welfare) (1979). *Smoking and health: Report of the Surgeon General.* Rockville, MD: Public Health Service, Office of the Assistant Secretary for Health, Office on Smoking & Health, DHEW Publication No. 79-50066.

USDHHS (U.S. Department of Health & Human Services) (1986). *The health consequences of involuntary smoking: Report of the Surgeon General.* Rockville, MD: Public Health Service, Office on Smoking & Health, DHHS Publication No. 87-8398.

USDHHS (U.S. Department of Health & Human Services) (1988). *The health consequences of smoking: Nicotine addiction: Report of the Surgeon General.* Rockville, MD: Public Health Service, Office on Smoking & Health, DHHS Publication No. 88-8406.

USDHHS (U.S. Department of Health & Human Services) (1989). *Reducing the health consequences of smoking: 25 Years of progress: Report of the Surgeon General.* Rockville, MD: Public Health Service, Centers for Disease Control Office on Smoking & Health, DHHS Publication NO. 89-8411.

USPHS (U.S. Public Health Service) (1964). *Smoking and health.* Report of the Advisory Committee to the Surgeon General of the Public Health Service (U.S. Department of Health, Education, and Welfare, Public Health Service, Health Services and Mental Health Administration, PHS Publication No. 1103). Washington, DC: U.S. Government Printing Office.

Vogler, G. P., & Fulker, D. W. (1983). Familial resemblance for educational attainment. *Behavior Genetics, 13,* 341–354.

CHAPTER FOUR

Smoking and Addictive Behaviors

Epidemiological, Individual, and Family Factors

DAVID C. ROWE AND MIRIAM R. LINVER

INTRODUCTION

This chapter deals with the acquisition among adolescents of smoking and related addictive behaviors (e.g., marijuana use). It is appropriate to focus on the adolescent period because few younger children have acquired smoking or other substance-use habits and because few people who do not acquire them during adolescence will do so later. Indeed, the "midlife crises" so readily diagnosed by pop psychology may be resolved by experimentation with adolescent-typical behaviors (e.g., a fast sports car), but not typically by experimenting with addictive substances. Adolescence thus qualifies empirically as a *critical period* for the acquisition of substance-use addictions. Nothing physically limits adults from acquiring such addictions, so it is more likely that the criticality of the adolescent period is a psychological phenomenon.

In Chapter 3, Heath and Madden review the evidence for genetic and environmental influences on the initiation and persistence of smoking. There appears to be a marked genetic contribution to smoking persistence that is not explained by correlations with personality or sociodemographic variables. For smoking initiation, the evidence again suggests that there are genetic influences, but in this case, in contrast to the results for persistence of the smoking habit, twin studies suggest that a substantial amount of variance is associated with environmental influences shared by twins. Moreover, adolescent twins who report that

---

DAVID C. ROWE AND MIRIAM R. LINVER • School of Family and Consumer Resources, University of Arizona, Tucson, Arizona 85721.

Behavior Genetic Approaches in Behavioral Medicine, edited by J. Rick Turner, Lon R. Cardon, and John K. Hewitt. Plenum Press, New York, 1995.

they have peer friends in common with their co-twins are more similar for smoking initiation than are twins who have separate peer-group friends. In this chapter, we offer a theoretical discussion and then a mathematical model to account for the acquisition of smoking in adolescence.

A good theory should integrate facts about substance-use addictions and generate new testable hypotheses. This biosocial theory of smoking will start with several accepted facts about the acquisition of smoking and related behaviors and attempt to explain them using biological and social concepts.

What facts must a biosocial theory of smoking account for? First, of course, is *who* will initiate smoking and later develop addictive habits. Not everyone (in any society) smokes cigarettes. These individual differences must be explained by a theory of smoking behavior. Second, the theory must explain *when* smoking habits are acquired—why in adolescence, not in other age periods. Third, the theory must account for the tendency of smoking habits to "run" in families. Why is the child of a smoking parent more likely to smoke (or to use other drugs) than other children? Fourth, the theory must explain the role of friends in the acquisition of smoking. Friendship groups tend to be homogeneous for smoking behavior—either smoking or not. Other facts about smoking also require explanation, e.g., the historical changes in smoking prevalence (i.e., declining in recent years) and an adolescent's movement from smoker to former smoker. The theory advanced here, however, cannot account for all facts about smoking; some issues, such as historical changes, will be touched on only lightly.

We take as our starting point the psychological meaning of smoking to young adolescents. Without understanding what cigarettes and related drugs offer psychologically to adolescents, one cannot begin to understand why they would be tried initially. Without a network of sales and distribution, though, adolescents would be unable to find and use cigarettes and other addictive substances; so an explanation of smoking behavior must take into account the social context of smoking. That is, the theory must incorporate both a psychological level of explanation—to explain the appeal of cigarettes—and a social level—to explain how adolescents obtain them.

With these ideas, the outline of the following argument about smoking can be presented in broad strokes. We will argue that cigarettes are used because, first, they are a *social signal,* like a raised flag, for adult sexual and social ambitions and, second, because they are pleasurable. Of the two psychological functions, we place greater emphasis on the former, because the pleasurable (and addictive) qualities of smoking would be discovered only once a large number of cigarettes have been smoked (initially, the effects of smoking and other drugs are often unpleasant).

To account for the familial aggregation of smoking, our theory emphasizes heritable personality traits that may affect the likelihood of experimenting with cigarettes (see Chapter 3). Such heritable traits would be related to the desire to break away from parents and to enter quickly into adult-status sexual and social behaviors. Rather than parents being social models of smoking, the tendency to smoke may be transmitted in families genetically through these heritable traits.

On the other hand, social influence on smoking may be mainly one of close-in-age peers. That is, smoking may be transmitted through social diffusion in adolescent peer groups, an idea that is given a mathematical representation later. Our theory postulates that peer-group opportunity to smoke is more important than peer influence to smoke, even though some adolescents are certainly pressured by friends into experimenting with cigarettes. In summary, this biosocial theory is one of biological dispositions being played out in an arena of socially constructed opportunities, which are enhanced by the social isolation of adolescents from adults in their somewhat separate society of the junior and senior high schools.

PSYCHOLOGICAL FUNCTIONS OF SMOKING

In the broadest terms, we assume that when behaviors persist and endure in human societies, they are likely to be biologically prepared.* On one hand, marriage rites exist in all societies and are handed down from one generation to the next, but these bonds also relate to human instincts arising from an evolutionary heritage. On the other hand, smoking behavior is a social innovation, introduced to European peoples within historical memory by American Indians (although other populations have smoked nontobacco substances) and then exported around the world. Nonetheless, although shorter in longevity than marriage rites, smoking has persisted in societies for hundreds of years; smoking behavior, too, may be supported by humans' biological dispositions.

Diamond (1992) extended an evolutionary explanation of animal display to include smoking and substance abuse. His argument followed from the evolutionary explanation by Zahavi (1975) of self-handicapping trait. In many animal species, display characteristics appear to be costly or self-destructive, e.g., a male peacock's long tail. The long tail makes it more difficult for the peacock to evade predators, but at the same time it is useful for attracting females. Zahavi theorized that such display traits had a general function: They may serve to advertise their bearer's ability to survive and overcome the "handicap" imposed by the display trait. Only an animal with unusually good traits overall would be able to carry the burden of the handicapping display trait. This display trait may attest to the genetic quality of its bearer. For example, Diamond (1992, p. 197) stated that

> . . . deleterious structures and behaviors constitute valid indicators that the signaling animal is being honest in its claim of superiority, precisely *because* those traits themselves impose handicaps [author's emphasis].

Diamond (1992) suggested that drinking and smoking may serve a similar signaling function in humans. He reasoned (p. 199):

> The messages of our old and new displays nevertheless remain the same: I'm strong and superior. Even to take drugs only once or twice, I must be strong enough to get past the burning, choking sensation of my first puff on a cigarette, or to get past the

*Persistence is an insufficient demonstration of a biological basis for behavior, of course. The use of fire has persisted for thousands of years, without its being clearly instinctive, except that fire provides a great many hedonistic benefits.

> misery of my first hangover. To do it chronically and remain alive and healthy, I must be superior (so I imagine). It's a message to our rivals, our peers, our prospective mates—and to ourselves.

Thus, smoking behavior would be a "message" of biological superiority intended for an audience of adolescent peers. This function would not necessarily require that smoking itself be biologically beneficial. As Diamond (1992) observed (p. 203):

> Drug addicts and drunkards not only lead shorter lives, but they lose rather than gain attractiveness in the eyes of potential mates and lose the ability to care for children. Those traits don't persist because of hidden advantages outweighing costs; they persist mainly because they are chemically addicting.

Indeed, the addictive properties of psychoactive drugs make their maintenance in the adult community rather easy to explain; addicted individuals find ceasing to use tobacco and other drugs extremely difficult. Once drug use reaches a given prevalence within a generation, this prevalence becomes self-maintaining, except through occasional quitting and deaths from its health consequences or overdoses (the latter being less significant for smoking than for other drugs).

Although the function of "proving biological superiority" might underlie the psychological attraction of smoking and other drugs, we believe that smoking is also used to advertise adult sexual and role status. The two ideas are not unrelated—smoking may indicate, "I'm an adult, capable of earning my own income and finding a mate," and indicate as well, "I'm tougher and stronger than other adolescents."

Smoking advertises adult status in several ways. First, smoking is a behavior clearly performed much more frequently by adults than by children. Second, the decision to smoke is one made by adolescents independently of adults. Given the current societal opposition to smoking behavior, a decision to smoke is all the more proof of independence from parents and adults; the adolescent is saying in effect, "I can make my own decisions." Third, advertisers work hard to associate smoking with adult virility. Cigarettes are seen being smoked by American icons like cowboys out on the range or by young, attractive, sophisticated urban women. In light of the known symbolic meaning of smoking, this behavior is naturally attractive to many adolescents.

Given our analysis, we would also expect smoking behavior to be associated with the biological basis of adult status: pubertal maturation. In our Arizona Sibling Relationship Study (Rowe & Gulley, 1992), pubertal maturation correlated 0.16 with smoking in boys and 0.22 with smoking in girls. Although these associations were not strong, they were limited by the high proportion of young (10–12 years old), pubertally immature children in the Arizona sample. Furthermore, it is almost impossible to know how much of this relationship is causally due to maturation, because physical maturation and age were inevitably highly correlated. Additional evidence for a biological connection comes from the direct relationship between blood testosterone levels and boys' and girls' smoking in early adolescence (Bauman, Foshee, Koch, Haley, & Downston, 1989). Testosterone levels also moderate social influences on smoking (Bauman, Foshee, & Haley, 1992). Friends' smoking had a greater relationship to respondents' smoking for boys with higher ($r = 0.52$) vs. lower ($r = 0.39$) testosterone levels. In girls, mothers' smoking was more strongly related to daughters' smoking when girls had a high ($r = 0.35$) vs. low ($r = -0.09$) testosterone level. The association of smoking onset with the adolescent life period thus appears to have a basis in biology, as suggested by our theoretical positions.

We would also expect smoking to be more prevalent among adolescents who want to try adult behaviors immediately than among those adolescents willing to delay adulthood for further education or skills acquisition. One test of this supposition is that sexually precocious adolescents should be more likely to smoke than sexually inexperienced ones.

We tested this prediction using data from the Arizona study mentioned above (Rowe & Gulley, 1992). The average age of the 836 teenagers in this study was 13.5 years. A sexual-experience scale was constructed from 5 items of progressively more intimate sexual behaviors, from kissing to sexual intercourse. Table 1 shows the results of the regression of sexual experience on age and smoking and its regression on age alone. The difference in the variance explained by these two regression equations is the contribution of smoking to sexual experience, independently of age. In boys, smoking accounted for 3% of the variance in sexual experience, independently of age. In girls, it accounted for 9% of sexual-experience variation, independently of age. The effects of smoking on sexual experience may be weakened in this sample because of its young age—most 10- to 12-year-old respondents were both nonsmokers and sexually inexperienced. Nonetheless, these results demonstrate that smoking behavior is an

TABLE 1. Smoking Behavior and Sexual Experience in Adolescents[a]

Regression	Variance explained in sexual experience in the last year
Males ($N = 411$)	
Age + smoking	0.42
Age alone	0.39
Smoking alone	0.03
Females ($N = 425$)	
Age + smoking	0.49
Age alone	0.40
Smoking alone	0.09

[a]This analysis is conservative if variance shared by age and sexual experience is causally due to sexual experience instead of age. All variance components are statistically significant ($p < 0.05$).

indicator of precocity of sexual experience—smokers tend to be more sexually advanced for their ages than nonsmokers.

Our theory thus far explains why some adolescents adopt smoking, but does not explain as well why some avoid doing so. If smoking were a powerful signal of superiority or adult status, why do not all adolescents take advantage of it, rather than, according to current national surveys of smoking behavior, only about one third? The fundamental reason is that the perceived advantages of smoking must be played off against its perceived costs. The psychological advantages of smoking are a claim of superiority (for which smoking's "costs" may be advantageous) and a signal of adult status. On the other hand, smoking carries with it dangerous health consequences including lung cancer, heart disease, and other disabilities, mostly delayed well beyond adolescence; it also carries the immediate disadvantage of bad taste, bad breath, tooth discoloration, and, for athletes, performance loss. Because many costs of smoking are delayed, the *only* adolescents who will weigh these costs will be those adolescents who plan for and are concerned about their distant (as opposed to immediate) futures. Slowly maturing adolescents may also weigh the costs more heavily because they are more psychologically mature when confronted with the decision of whether to smoke. On the other hand, rapidly maturing adolescents, who are psychologically immature when confronted with the decision to smoke, should be more disposed to experiment with cigarettes, as would be adolescents who want an early entry into adult social behaviors. These latter individuals may be more impulsive and less self-controlled, or they may perceive less opportunity for scholastic success. This outline suggests the existence of a set of personality traits (e.g., impulsivity/self-control and intelligence) that may predict the onset of smoking. In a later section, such personality-trait correlates of smoking will be considered.

Our biosocial theory offers an explanation for historical changes in smoking prevalence. Prior to World War II, smoking was largely a male habit. During the war, many young women took factory jobs held previously by men. At the same time, smoking became popular among these women. Smoking could have served as a signal of precocious adult behaviors among women prior to 1945 as well as after, but it did not. We can speculate that as young women moved into previously restricted work roles during World War II, they decided to experiment with smoking, making what was previously a marker of male adult status their own. Once a high proportion of women became addicted, of course, smoking was able to maintain itself in the female population, as it had previously in the male. Among white Americans aged 18–24 years, the proportion of male and female regular smokers was nearly equal in 1985 (about 30%) (Rogers, 1991). It is noteworthy that, unlike criminal forms of social deviance—which maintain large rate differences between the sexes—prevalence rates for smoking and marijuana use are nearly equal for males and females. In this way, drug use is quite distinct from other types of behavioral deviance that involve victimizing others, or putting oneself in some physical danger, traits more associated with males than with females. Even though the sexes are about equal in drug-use levels, however, the marketing and distribution of illicit drugs remains predominantly a male activity.

The assumed psychological function of smoking does not necessarily apply to all societies and circumstances. In the recent civil war in the former Yugoslavia, cigarettes have become a commodity as valuable as hard cash: A single pack may sell for as much as $50. The civilians in besieged cities have discovered that cigarettes provide relief from the monotony of war and from hunger; hence, they are very much in demand. Nonsmokers have started to smoke just for relief from the terrible pressures of a war waged against civilians, in Sarajevo and in other Bosnian cities. The long-term ill-health effects of smoking, in this context, are regarded as a rude joke. These "adult-onset" cases of smoking fall outside the pattern dictated by our theory; they reflect the enormously different social circumstances in a war-torn country as opposed to the United States or the Western European democracies.

PERSONALITY TRAITS AND SMOKING

The theory described above revealed several traits with possible relationships to smoking behavior. They can be placed within the context of the "Big-Five" factor model of personality. On the basis of extensive factor analyses of self-reports and personality ratings, five distinct dimensions of adult personality have been identified: (1) extraversion, (2) agreeableness, (3) conscientiousness, (4) emotional stability, and (5) culture (Norman, 1963; Loehlin, 1992). Of these trait dimensions, the second one, conscientiousness, may have the strongest relationship with smoking behavior, a relationship that would be consistent with the association of smoking initiation and Cloninger's Novelty Seeking dimension reported in Chapter 3. Cloninger (personal communication) found that the "Big-Five" trait of conscientiousness correlated −0.48 with Novelty Seeking in a sample of 803 college students and −0.41 in a sample of 136 psychiatric patients; this empirical relationship is not surprising, since both smoking initiation and novelty seeking subsume impulsivity as one of their components. Individuals who are high in conscientiousness are reliable, organized, and persistent; those who are low in conscientiousness are impulsive, unreliable, and disorganized. As teenagers, individuals who are low in conscientiousness may be more motivated to enter immediately into adult behaviors—to show independence from their birth families and to become sexually active. In contrast, individuals high in conscientiousness may avoid adult behaviors and thus may delay initiating smoking. As noted above, if these high-conscientiousness individuals also fear the social risks incurred by smoking, then they may never initiate smoking, or if they do experiment with cigarettes, they may stop before physical and psychological addiction can take place.

A computer search located 46 studies on smoking published since 1986. Although no study adopted a "Big-Five" approach to the etiology of smoking, several examined traits related to conscientiousness. Three studies reported a positive association between sensation-seeking and smoking behavior (Chassin, Presson, & Sherman, 1989; Mosbach & Leventhal, 1988; Collins et al., 1987). In one study (Heaven, 1989), no association was found between smoking and the Big-Five dimensions of extraversion and neuroticism. One study also found that smokers were more "deviant" in their higher prevalence of petty crimes and in their greater tolerance of deviance (Chassin et al., 1989). As these results were on the basis of smoking transitions, they support a causal effect of prior behavioral deviance on subsequent smoking behavior. Viewing the studies overall, though, we got the impression that personality was not assessed using highly reliable scales or diverse sets of traits. Furthermore, the proportions of variance in smoking explained by personality were low (<10%). While our biosocial approach receives some support from these data, additional work defining the trait domain most strongly associated with smoking behavior is needed.

INTERGENERATIONAL TRANSMISSION OF SMOKING

We know that children are not born smokers; they must acquire smoking habits via cultural opportunities. Two models of intergenerational transmission predominate in the smoking literature. One is the vertical model of transmission, i.e., from parent to child. Of course, this possibility is limited to societies in which smoking is already established in the adult generation; it cannot explain the entry and spread of smoking in a nonsmoking society (today, few opportunities remain to observe this latter transition empirically). The horizontal model focuses on the transmittal of smoking from smoking peers, i.e., from individuals close in age to the nonsmoker. Each "generation" then initiates the next "generation"—just a few years younger—into the smoking habit.

Vertical Transmission

Being a child of a smoking parent increases the risk of smoking behavior. According to our theory, however, adolescent children would initiate smoking for reasons different from their parents' maintenance of smoking. The parent smokes because of physical and psychological dependence on nicotine in cigarettes as a psychoactive drug. The adolescent child initiates smoking as a way of demonstrating dominance over a self-inflicted handicap and as a way of signaling early initiation into adult behaviors. Children with traits that lead them away from early initiation of adult roles have little reason to initiate smoking, even if their parents smoke. In the United States today, many smoking parents also actively discourage smoking in their children, because they accept that cigarettes have harmful effects on health. Children with the personality and social traits detailed earlier that lead them toward smoking should be motivated to initiate smoking behavior regardless of parental advice. In our theory, the characteristics of the child, not those of the parents, are crucial to the initiation and early maintenance of smoking behavior; thus, we postulate very little direct social influence of parents on their children if their children were not already predisposed toward smoking. As will be discussed later, even in this case, parents are only one of many available social models.

How, then, does one explain familial risk for smoking behavior? The answer is that parents and children possess some personality resemblance because personality traits are heritable. Quantitatively, parent–child smoking correlations are quite weak. The typical parent–child correlation is only 0.10–0.20. Such a low correlation coefficient is consistent with a personality explanation of parent–child resemblance in smoking behavior. Although the total heritability of most personality traits is about 0.50, their additive heritability—the part responsible for parent–child personality resemblance—is only about 0.30 (Rowe, 1994). Thus, we assume that a composite of personality traits creates an increased risk of smoking behavior—a composite with a heritability of about 0.30. If the parent–child correlation for smoking were based on this composite, it would be only about 0.15 (i.e., because parents and children share only half their polymorphic genes, the parent–child correlation is $^1/_2 h^2$). These values are reasonable in light of results from different studies of smoking behavior.

Of course, a social influence model could also account for a low parent–child correlation by postulating that many extrafamilial influences buffer the child against parental example. To decide this issue, research designs are required that control for genetic influence while estimating environmental effects. Eysenck (1980) conducted a mail-survey family and adoption study of smoking when both the "children" were young adults. In the biologically related pairs, he found that smoking by a parent correlated 0.21 with that by the (adult-age) child. His survey thus replicated the usual familial-risk finding in biological families. In adoptive families, however, the correlation of parental and child smoking was statistically nonsignificant and close to zero in magnitude ($r = -0.02$). As expected under a genetic transmission model of smoking behavior, when parent–child genetic similarity is largely absent in the adoptive home, so is an association on smoking behavior.

A large-scale twin study confirms this finding using a different research design. In a twin study, effects of shared-rearing environments must be inferred indirectly from a statistical model of twin correlations. If fraternal or dizygotic (DZ) twins correlate on a trait as much as identical or monozygotic (MZ) twins, then (under certain assumptions) (see Plomin, 1986) shared rearing influences, but not genetic ones, may influence the trait. When identical twins are more alike in behavior than fraternal twins, then genetic influence can be inferred, because identical twins share 100% of genes relevant to trait variation, whereas fraternal twins share only (on average) 50% of the same genes. But this 2:1 ratio of genetic similarity has a further, quantitative implication: When a trait is determined wholly by genetic influences and environmental ones not tied to the family unit, the MZ and DZ twin correlations should possess a 2:1 ratio. Thus, a trait

TABLE 2. Twin Correlations for Cigarettes per Day and Parameter Estimates[a]

r		Heritability	Shared environment
MZ twins	DZ twins	(h^2)	(c^2)
0.42	0.18	0.48	−0.06

[a]For computational formulas, see Rowe (1994).

like height that is genetically determined with little family environment effect produces correlations on the order of 0.85 and 0.425, a 2 : 1 ratio. The same ratio may obtain for a more weakly genetic trait—e.g., when $r_{MZ} = 0.18$, $r_{DZ} = 0.09$—but the same inference applies, namely, that the environmental influences are not tied to the family.

As reviewed in Chapter Three, twin data support our inference from the findings with adoptive families. For example, one study (Swan, Carmelli, Rosenman, Fabsitz, & Christian, 1990) initially recruited twins through a twin registry of United States war veterans. Twin type was determined from 22 polymorphic gene loci. At the average age of 59 years, 360 pairs (176 MZ and 184 DZ) were located again, and they reported their own cigarette-smoking consumption. As shown in Table 2, the twin correlations for smoking (cigarettes/day) were 0.42 and 0.18 for MZ and DZ twins, respectively. In calculating these correlations, smoking rates were adjusted for variance shared with drinking and coffee consumption. The heritability estimate was 0.48; that of shared-rearing influences was −0.06 (nonsignificant). The heritability estimate obtained in this study for smoking behavior was close to the 0.53 average reported in the review by Hughes (1986) of the genetics of smoking and well within the ranges reported by Heath and Madden in Chapter Three. As in Eysenck's adoption study, no evidence was found for a shared family-rearing influence on smoking behavior.

Of course, both Eysenck's study and Swan and his colleagues' twin study dealt with older samples. It is possible that there is some family environmental influence on the initiation of smoking in adolescence (see Chapter 3), but this influence diminishes as adolescents either continue to smoke as adults or quit smoking. Our view is that the direct effects of family environment on smoking behavior, even at younger ages, would be small because peer influence dominates in the initiation of smoking. Nonetheless, data on the smoking behavior of adolescent adoptees raised by smoking and nonsmoking adoptive parents would serve to strengthen our theory.

Horizontal Transmission

In the United States and in Western European democracies, schooling is a universal experience for young children and adolescents. Except for their contact with teachers, school administrators, and other school personnel, children and teenagers are isolated from adult society for a major part of their waking hours. Some historians attribute the development of the concept of adolescence itself to sequestering young people from adults and allowing them to form, in a sense, a separate society. Of course, adolescence has characteristics that predated universal schooling—most notably, its inevitable association with pubertal maturation. Nevertheless, the age-stratification of the United States and Western European adolescents has been less permeable than that in most preindustrial states or that in tribal societies (because adolescents are segregated within school systems by grade level). The physical structure of adolescent society thus can encourage a diffusion of behaviors among peers.

Given our view that the psychosocial function of smoking is to make a social statement (primarily) to one's adolescent peers, and given this age-graded structure of the adolescent society, one might expect that social meetings with peers are crucial for the initiation of smoking behavior. This expectation receives support from survey data on the onset of smoking. According to in-depth interviews with 157 adolescents (12–19 years old), the majority initiated smoking with another adolescent close in age (friend, acquaintance, or sibling, 80%) (Friedman, Lichtenstein, & Biglan, 1985). Only 3% reported initiating with a parent. In this group, 11% reported that they were alone. Other small categories were relative (5%) and stranger (4%).* While 70% of adolescents claimed they were not *pressured* by the other individuals present, the majority (76%) also reported that one or more of those individuals suggested smoking; 63% also said that another person offered them a cigarette. On the other hand, 22% of those polled said that they were planning to smoke at the time. Smokers and nonsmokers in these groups were primarily same-sex individuals. The overall picture is one of an intimate small group, in which mild social encouragement to smoke may have been given and in which some adolescents were already prepared to begin smoking.

These social conditions of smoking initiation led us to model it by analogy to an epidemic of infectious disease (Rowe & Rodgers, 1991; Rowe, Chassin, Presson, Edwards, & Sherman, 1992). Of course, smoking behavior is not a disease. This analogy with epidemic diseases is not meant to imply that smoking is either a sickness or somehow immoral (despite opposition, smoking remains a legal behavior in the United States). Rather, the mathematics of infectious diseases (May & Anderson, 1987) and social innovations (Mahajan & Peterson, 1985) follow one set of mathematical laws that describe the spread of an infectious agent or a new behavior through face-to-face encounters (Rodgers & Rowe, 1993). In this application, this new innovation is smoking behavior, which is rediscovered by each generation of adolescents. Thus, it is much like any other social innovation; its success depends on the attractiveness or positive utility that people attach to it (the psychological utility of smoking was described above). The greater the psychological utility, the more rapidly smoking will spread through social networks of adolescents.

*The percentages do not total 100% because a 3% discrepancy existed between the questions "Were you alone? and "Who was present?"

In our first mathematical model of smoking (Rowe et al., 1992), we attempted to evaluate our theoretical ideas about the role of peer transmission. The term *peer contagion* implies an active effort by a smoking peer to influence the smoking behavior of another peer. The term *peer diffusion* refers to a more passive process—an adolescent is with a peer who is smoking and decides to try it. Our model is neutral about the degree of influence or diffusion present in the typical encounter between a smoker and a nonsmoker. Both processes are presented in the model by a process of prevalence-driven transition between smoking stages.

The term *Prevalence-driven* merely means that the rate of transition depends on the prevalence of smoking in a population. The population is assumed to "mix"; i.e., smokers will encounter both other smokers and nonsmokers. These mathematical models treat the mixing process as a random one—clearly, an oversimplification of the social world because smokers tend to socialize preferentially with smokers, and vice versa (Fisher & Bauman, 1988). The existence of a bias of "like socialize with like" would slow the social spread of smoking behavior (i.e., because of the lost opportunities of dissimilar people to mix). However, in a simulation analysis, we found that parameter estimates of epidemic models were fairly robust to violations of the random-mixing assumption (Rowe, Rodgers, & Meseck-Bushey, 1989). Furthermore, adolescent friendships and social cliques tend to be short-lived, encouraging considerable "mixing" within adolescent populations. In lieu of computer simulation models that can relax this assumption, we make use of this assumption in our mathematical "epidemic" models.

In a randomly mixing population, a prevalence-driven transition takes the form

$$Ptr_{a+1} = TPns_a M_a \tag{1}$$

where Ptr_{a+1} is the proportion of experimental smokers (triers) of age a + 1, T is a rate constant, Pns_a is the proportion of nonsmokers of age a, and M_a is the proportion of smokers of all types (i.e., both triers and regular smokers) of age a. The rate at which nonsmokers convert to smokers thus depends on the prevalence of adolescents who already smoke (M in equation [1]).

Not all transitions demand social contacts, however. In the epidemic model of smoking, four "stages" of smoking behavior were identified: nonsmoker, trier (smokes just a few times), regular smoker (smokes weekly or more), and former smoker. Figure 1 displays these smoking stages. An adolescent is assumed to progress stage by stage; i.e., to jump from nonsmoker directly to regular smoker would violate the model. Using longitudinal data, we verified that most adolescents follow the stage progression in Figure 1 (Rowe et al., 1992). We proposed that the transition from nonsmoker to trier was prevalence-driven, as described above in equation (1). However, we thought that the transition from trier to regular smoker would not depend so much on social contacts as on the psycho-

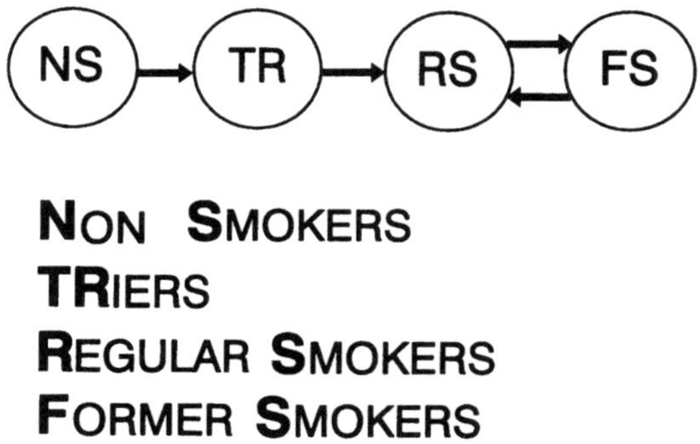

FIGURE 1. Smoking transitions permitted in the epidemic model. Reprinted by permission from *The Limits of Family Influence: Genes, Experience, and Behavior*, by David C. Rowe, p. 205. Copyright 1994 by Guilford Press.

logical and physiological rewards of cigarettes themselves. On the basis of this premise, transition from trier to regular smoker was expected to occur at a constant rate.

Mathematically, a constant rate transition means that a constant proportion of adolescents makes a transition in a particular period. In this case

$$Prs_{a+1} = jPtr_a \tag{2}$$

where Prs_{a+1} is the proportion of regular smokers of age a + 1, j is a rate constant, and Ptr_a is the proportion of triers of age a.

The complete model required four equations, one each to represent the proportion of nonsmokers, triers, regular smokers, and former smokers, respectively, of each age. The proportion of nonsmokers of age a + 1 was expressed in terms of the proportion of nonsmokers of the prior age and the loss due to nonsmokers who became smokers. The following equation shows this process:

$$Pns_{a+1} = Pns_a - TPns_a\mathrm{M}_a \tag{3}$$

where the first two terms were previously defined, T is a rate constant, and M_a is the sum of the proportions of regular smokers plus triers of age a.

The next equation was derived for the proportion of triers (*Ptr*) as follows:

$$Ptr_{a+1} = Ptr_a + TPns_aM_a - jPtr_a \tag{4}$$

where all terms were previously defined. In words, triers at age a + 1 equal old triers plus nonsmokers who adopt smoking behavior minus triers who become regular smokers. The last group is added to the proportion of regular smokers in the following equation:

$$Prs_{a+1} = Prs_a + jPtr_a - uPrs_a + vPfs_a \quad (5)$$

where *Prs* is the proportion of regular smokers, *u* is the quitting rate, and *v* is the relapse rate of former smokers, Pfs_a. Finally, the cycle of transitions is completed by the following equation for the proportion of ex-smokers:

$$Pfs_{a+1} = L - Pns_{a+1} - Ptr_{a+1} - Prs_{a+1} \quad (6)$$

where *L* is the proportion of nonimmune adolescents. An immune status was introduced because adult follow-up data indicate that some individuals never leave the nonsmoker status (i.e., they never experiment with cigarettes during adolescence). The immune class was defined on the basis of the proportion of individuals who never tried smoking in the longitudinal follow-up of adolescents. For instance, if 15% of the sample were immune, and 45% were in the otherwise susceptible class in the first year of observation, then this 45% would be reduced to 30% to guarantee that the initial susceptible group included only individuals who later progressed through one or more smoking stages (for details, see Rowe et al., 1992). Of course, other models would be possible, including ones ignoring the concept of "immunity" altogether.

The equations were fitted to the smoking behavior of 6th- to 11th-graders in a longitudinal study of smoking. For our smoking variables, the sample size ranged from 4250 to 4624. Smoking was measured with a single questionnaire item that divided the adolescents into the four categories. In this study, a fairly liberal definition of regular smoking was used (smokes monthly or more), whereas trier was strictly defined (smoked just one or two times). Former smokers identified themselves as having quit; nonsmokers, of course, reported no use of cigarettes. Use of chewing tobacco or snuff was not considered. The validity of self-report measures of smoking cigarettes is generally very good against biological or informant corroboration.

Several interesting findings emerged from the application of the "epidemic" model. First, the best fits of the model to the data were obtained when the transition from nonsmoker to trier involved social contacts. That is, modifying the equation so that the term $TPns_aM_a$ became $TPns_a$ worsened fit. As can be seen, the latter term omits social contacts with smokers, represented mathematically as the proportion of smokers of all types, M_a. Second, parameter estimates for females and males were similar, with T = about 0.71 and 0.60, respectively, in the two sexes. The transition between trier and regular smoker, however, was about twice as rapid in females as in males ($j = 0.16$ and 0.09, respectively). This similarity of the *T* parameters suggests that females and males are not completely separable populations (which would lead to a divergence of contagion rates), but may "cross-infect" one another frequently enough to tend to equalize the social contagion effect. Finally, the χ^2 values assessing the inconsistency between the model and the data were usually significant, but not in all four "waves" of data on the sample. The model fit the data collected in 1980 best, with χ^2 equal to 14 in females and 21 in males, for 12 degrees of freedom.

A single mathematical model cannot capture all the complexities that underlie the initiation and maintenance of smoking. For instance, the interpretation of the single number *T*, the rate parameter for transition between nonsmoker and trier, is quite complex. The number represents both opportunities and individuals' responses. Opportunities are the raw rate of smoker–nonsmoker social contacts in situations conducive to acquiring smoking behaviors; individuals' responses are their willingness to start smoking, given the opportunity to do so. Moreover, the number hides the population heterogeneity that is at the heart of individual variation. The rate of transition, *T*, is not the same for all adolescents. For those thought to be immune to social pressures to smoke, it is close to zero. For an adolescent who is low in conscientiousness, *T* may be high in value. Models working with aggregate data may conceal the importance of such individual differences.

There are several ways of uniting the "epidemic" approach with individual-level analyses. One method is to investigate "epidemic processes" in different populations, e.g., in the children of smoking parents vs. nonsmoking parents. Our preliminary results suggest that the *T* parameter in our epidemic model is especially influenced by parental smoking status. That is, transitions from nonsmoker to trier occurred considerably more often in the adolescent children of smoking compared with nonsmoking parents. Of course, subdividing a population as a means of introducing a covariate is crude; it is also limited to a small number of covariates, lest cells in the experimental design be empty. To extend the model to encompass individual differences fully, we are currently pursuing computer simulation methods to find more accurate representations of the spread of smoking behavior among adolescents.

SUMMARY AND CONCLUSIONS

Our perspective on smoking behavior starts with the psychological meaning of smoking to adolescents. It assumes a prevalent social stereotype of smokers among adolescents such that smoking behavior itself becomes a social signal, like hair or dress style, to other adolescents, in addition to its pleasurable qualities. For both sexes, the stereotype may involve rapid maturation, a desire to break bonds with birth families and move quickly into adult social behaviors such as sexuality and decisional independence. For males especially, the behavior may also convey a toughness—not unlike the apparently "handicapping" traits of some animals that may convey genetic quality because a visible handicap must be overcome. Unfortunately, few published studies bear directly on the social meaning of smoking behavior to adolescents. Thus, one recommendation is for work focusing on the social meaning of smoking.

Given its likely social meaning, our perspective further asserts that individual variation in proneness to smoke is related to broad personality traits, in particular to the "Big-Five" personality dimension of conscientiousness. The argument assumes that individuals low in conscientiousness would be more impulsive, less oriented toward long-term outcomes, less sensitive to punishment, and hence more likely to engage hedonistically in behaviors that lead to immediate psychological and physical pleasure. Here, too, the data were less than adequate to the task. Few studies of smoking employed a broad conception of personality with measures that had adequate statistical reliability; our inference that the trait of conscientiousness is a key determinant of smoking proneness thus depends on our classification of the traits for which data were available belonging to this dimension rather than to another. Thus, further studies are needed on the relationship of the "Big-Five" personality dimensions with smoking behavior.

In terms of the intergenerational and intragenerational transmission of smoking, our perspective emphasized the "horizontal" transmission of smoking behavior within a generation over the "vertical" transmission of smoking between parent and child. Although familial resemblance exists for smoking behavior, it is interpreted here as an association induced by parent–child genetic resemblance rather than one determined by parental modeling of smoking behavior or by some other social environmental effect. The main support for this conclusion was the absence of consistent shared environmental variation in smoking behavior. Nonetheless, the behavior genetic studies cited in this chapter and Chapter Three used primarily adult respondents. It is possible that parents exert on their children's smoking behavior short-term environmental effects that later dissipate, a view that is also consistent with the finding that where shared environmental variation is indicated, it is for smoking *initiation* rather than for persistence (see Chapter Three).

In our view, however, the social transmission of smoking is likely to be primarily horizontal, i.e., among adolescent peers, rather than vertical from parents to children. Accordingly, an epidemic model was detailed that mathematically describes a process of social spread. Of course, the model was mute concerning the degree of "influence" vs. "simple diffusion" (i.e., opportunity). Although friends' smoking behaviors are clearly statistically associated, not all this association is attributable to the influence of one adolescent on another. Some adolescents are ready to accept smoking behavior, with little "push" to do so. An elegant longitudinal study indicates that "peer pressure" is probably only a small component of friends' resemblance in smoking—with the majority of the association arising from the tendency of individuals who smoke to befriend one another (Fisher & Bauman, 1988).

In summary, smoking behavior appears to be transmitted through the adolescent population by the contact of smokers with already susceptible individuals; some peer pressure may be applied, but it is not totally opposed by the nonsmoker who is susceptible. Later on, maintenance of smoking in adults may depend more on the psychologically and physically addictive qualities of nicotine in cigarettes. Our perspective thus gives different explanations for early initiation of smoking behavior than for long-term smoking in adulthood and is in broad agreement with the empirical findings reviewed in Chapter Three. The theory we have outlined combines constitutional and social components in an approach that we hope will contribute to further advances in understanding smoking behavior.

REFERENCES

Bauman, K. E., Foshee, V. A., & Haley, N. J. (1992). The interaction of sociological and biological factors in adolescent cigarette smoking. *Addictive Behaviors, 17,* 459–467.

Bauman, K. E., Foshee, V. A., Koch, G. G., Haley, N. J., & Downton, M. I. (1989). Testosterone and cigarette smoking in early adolescence. *Journal of Behavioral Medicine, 12,* 425–433.

Chassin, L., Presson, C., & Sherman, S. (1989). "Constructive" vs. "destructive" deviance in adolescent health-related behaviors. *Youth and Adolescence, 18,* 245–262.

Collins, L., Sussman, S., Rauch, J., Dent, C., Johnson, C., Hansen, W., & Flay, B. (1987). Psychosocial predictors of young adolescent cigarette smoking: A sixteen-month, three-wave longitudinal study. *Journal of Applied Social Psychology, 17,* 554–573.

Diamond, J. (1992). *The third chimpanzee: The evolution and future of the human animal.* New York: HarperCollins.

Eysenck, H. (1980). *The causes and effects of smoking.* London: MT Smith.

Fisher, L. A., & Bauman, K. E. (1988). Influence and selection in the friend–adolescent relationship: Findings from studies of adolescent smoking and drinking. *Journal of Applied Social Psychology, 18,* 289–314.

Friedman, L. S., Lichtenstein, E., & Biglan, A. (1985). Smoking onset among teens: An empirical analysis of initial situations. *Addictive Behaviors, 10,* 1–13.

Heaven, P. (1989). Adolescent smoking, toughmindedness, and attitudes toward authority. *Australian Psychologist, 24,* 27–35.

Hughes, J. R. (1986). Genetics of smoking: A brief review. *Behavior Therapy, 17,* 335–345.

Loehlin, J. C. (1992). *Genes and environment in personality development.* Newbury Park, CA: Sage Publications.

Mahajan, V., & Peterson, R. A. (1985). *Models for innovation diffusion.* Beverly Hills, CA: Sage Publications.

May, R. M., & Anderson, R. M. (1987). Transmission dynamics of HIV infection. *Nature, 326,* 137–142.

Mosbach, P., & Leventhal, H. (1988). Peer group identification and smoking: Implications for intervention. *Journal of Abnormal Psychology, 97,* 238–245.

Norman, W. T. (1963). Toward an adequate taxonomy of personality attributes: Replicated factor structure in peer nominations personality ratings. *Journal of Abnormal and Social Psychology, 66,* 574–583.

Plomin, R. (1986). *Development, genetics, and psychology.* Hillsdale, NJ: Lawrence Erlbaum Associates.

Rodgers, J. L., & Rowe, D. C. (1993). Social contagion and adolescent sexual behavior: A developmental EMOSA model. *Psychological Review, 100,* 479–510.

Rogers, R. G. (1991). Demographic characteristics of cigarette smokers in the United States. *Social Biology, 38,* 1–12.

Rowe, D. C. (1994). *The limits of family influence: Genes, experience, and behavior.* Guilford Press.

Rowe, D. C., Chassin, L., Presson, C. C., Edwards, D., & Sherman, S. J. (1992). An "epidemic" model of adolescent cigarette smoking. *Journal of Applied Social Psychology, 22,* 261–285.

Rowe, D. C., & Gulley, B. L. (1992). Sibling effects on substance use and delinquency. *Criminology, 30,* 217–233.

Rowe, D. C., & Rodgers, J. L. (1991). Adolescent smoking and drinking: Are they "epidemics"? *Journal of Studies in Alcohol, 52,* 110–117.

Rowe, D. C., Rodgers, J. L., & Meseck-Bushey, S. (1989). An "epidemic" model of sexual intercourse prevalences for black and white adolescents. *Social Biology, 36,* 127–145.

Swan, G. E., Carmelli, D., Rosenman, R. H., Fabsitz, R. R., & Christian, J. C. (1990). Smoking and alcohol consumption in adult male twins: Genetic heritability and shared environmental influences. *Journal of Substance Abuse, 2,* 39–50.

Zahavi, A. (1975). Mate selection—a selection for a handicap. *Journal of Theoretical Biology,* 53, 205–214.

PART THREE

STRESS AND REACTIVITY

CHAPTER FIVE

Behavior Genetic Studies of Cardiovascular Responses to Stress

JOHN K. HEWITT AND J. RICK TURNER

INTRODUCTION

Systematic research in the fields of cardiovascular psychophysiology, behavioral medicine, and psychosomatic medicine has revealed that there is considerable individual variation in psychophysiological responses to psychological stress (Manuck, Kasprowicz, Monroe, Larkin, & Kaplan, 1989; Obrist, 1981; Turner, 1989). While all individual difference phenomena are of intrinsic interest to behavioral scientists, variation in cardiovascular stress responses has attracted additional attention as a result of hypothesized links between large stress responses and the later development of cardiovascular disease (Blascovich & Katkin, 1993; Matthews et al., 1986; Manuck, 1994; Turner, Sherwood, & Light, 1992).

The extent of the pathogenic influence of stress-induced cardiovascular responses depends in part on those responses' reflecting stable and consistent individual differences—in other words, on their showing reproducibility across time and across different stressors (Matthews, Rakaczky, Stoney, & Manuck, 1987). Accumulating evidence suggests that there is a reasonable degree of both temporal and intertask consistency for such stress responses (Sherwood and Turner, 1992; Turner, 1994), and it is therefore appropriate to investigate the

---

JOHN K. HEWITT • Institute for Behavioral Genetics, University of Colorado at Boulder, Boulder, Colorado 80309. J. RICK TURNER • Departments of Pediatrics and Preventive Medicine, University of Tennessee, Memphis, Tennessee 38105; *present address:* Department of Pediatrics and Georgia Prevention Institute, Medical College of Georgia, Augusta, Georgia 30912.

Behavior Genetic Approaches in Behavioral Medicine, edited by J. Rick Turner, Lon R. Cardon, and John K. Hewitt. Plenum Press, New York, 1995.

origins of their differences among individuals. In this chapter, we review empirical evidence for genetic influences on these responses.

As is evident throughout this book, one of the most powerful, practicable, and versatile ways to explore the roles of genetic and environmental determinants of individual variation is the twin study. We therefore review the reported twin studies of cardiovascular responses to laboratory psychological challenge.

We then outline what we consider to be fruitful avenues for future research on the quantitative genetics of cardiovascular stress responses. We pay particular attention to the promise offered by studies of ambulatory cardiovascular profiles, which combine the methodology of behavior genetics with recently available technology for ambulatory monitoring.

LABORATORY INVESTIGATION OF STRESS RESPONSES

With regard to psychosomatic theories of cardiovascular disease, ultimate interest lies with physiological responses to the stresses of everyday living in our natural environments. If large stress responses play a role in the etiology of pathophysiology, it is in the arena of real-world challenge and everyday psychosocial interactions that these responses will take their toll (Turner, 1989). Nevertheless, much useful information can be gathered in the laboratory by employing a psychophysiological strategy. Cardiovascular psychophysiologists have therefore devised a range of laboratory stressors that elicit considerable physiological responses in certain individuals; these stressors include the Stroop color-word test, video games, reaction time tasks, mental arithmetic, speech tasks, and mirror drawing tasks. It is tasks such as these that have largely been used in the laboratory investigation of genetic influence on stress responses.

While this chapter will focus on psychological stressors, it should be noted at this point that the cold pressor task (see Lovallo, 1975; Saab et al., 1993) has a venerable history in cardiovascular research. For example, and of particular relevance here, McIlhaney and Hines examined the responses of 200 pairs of young twins to the cold pressor. Following a preliminary report (Hines, McIlhaney, & Gage, 1957), a subsequent paper presented evidence of genetic influence on blood pressure responses to the task (McIlhaney, Shaffer, & Hines, 1975). Accordingly, it is fitting to acknowledge the study of McIlhaney and Hines here, since it is perhaps the earliest twin study of cardiovascular responses to a stressor (Rose, 1986).

Twin Studies of Stress Responses

In this section, we review ten published studies that have explored the origins of individual differences in cardiovascular response to psychological challenge by employing the classic twin study methodology. These studies focus on heart rate or blood pressure or both as indices of cardiovascular activity.

(Additional relevant studies, such as Boomsma (1992), are presented in Chapter Six.) In an attempt to facilitate interstudy comparison and to generate overall conclusions, we provide statements of the range of heritabilities with which the reported results are consistent. However, we emphasize that these values are to be considered only ballpark figures: On only three of the ten occasions were formal estimates of heritabilities provided together with their standard errors or with adequately informative statistical tests of their significance.

Vandenberg, Clark, and Samuels (1965) employed 38 pairs of twins—22 pairs of monozygotic (MZ) twins and 16 pairs of dizygotic (DZ) twins—all high school students who were part of the Hereditary Abilities Study at the University of Michigan. Mild stress was administered in the form of startle stimuli: a flash of light, the ringing of a doorbell, and the falling of a hammer. Heart rate responses were calculated by subtracting the resting rate from that after each stimulus. Within-pair variance of DZ twins was compared with that for the MZ twins by means of the F ratio. Significant F ratios, indicative of genetic variation, were obtained for responses to the light flash and hammer stimuli, and while heritability estimates were not reported, the results were consistent with heritabilities in the 70–80% range.

Shapiro, Nicotero, and Scheib (1968) exposed 24 twin pairs (12 MZ, 12 DZ) to the Stroop color-word test while monitoring heart rate and mean arterial pressure. Intraclass correlations that can be computed from the data presented by these authors are consistent with a heritability estimate of around 30% for heart rate responses to the stressor and one of 70% for blood pressure responses (see Turner and Hewitt, 1992).

Theorell, deFaire, Schalling, Adamson, and Askevold (1979) reported a study in which twin pairs (17 MZ, 13 DZ) were given a structured personal interview. Heart rate and systolic and diastolic blood pressure were recorded. Analysis was conducted for resting levels, task levels, and the task/rest ratio. Results were consistent with heritability in the 80–90% range for the task levels, but there was no evidence of genetic influence when the task/rest ratio was examined.

In a study conducted by Rose, Grim, and Miller (1984), subjects completed three psychological stressors: mental arithmetic, the Stroop test, and mirror drawing. The authors presented representative patterns of systolic pressure during the tasks, one for a pair of MZ and one for a pair of DZ twins. The similarity in level and pattern of blood pressure in the MZ twins contrasted with the discordant profiles of the DZ pair.

While the four studies discussed above certainly provide some pertinent information, they employed either relatively few subjects or analytical techniques that were not powerful enough to afford clear conclusions. The next study to be discussed employed a somewhat larger, balanced sample of MZ and DZ twins and utilized a more powerful test of alternative genetic and environmental models of variation afforded by weighted least-squares model-fitting. Carroll, Hewitt, Last, Turner, and Sims (1985) conducted a study in which 40 MZ and 40

DZ male twin pairs completed a video game task. Heart rate responses during the task were assessed. Preliminary correlation analysis was followed by weighted least-squares model-fitting (see Eaves, Last, Young, & Martin, 1978). A model postulating additive genetic effects, along with those arising from individual environments, best fit the data. A heritability estimate of 0.48 ± 0.11 was calculated for heart rate reactivity.

Turner, Carroll, Sims, Hewitt, and Kelly (1986) reported further data from an extension of this twin study. On this occasion, male twin pairs (22 MZ, 29 DZ) completed two tasks: the video game used in the Carroll et al. (1985) study and a mental arithmetic task. Heritability estimates for heart rate responses were 0.61 ± 0.12 and 0.55 ± 0.13, respectively, for the video game and arithmetic stressors.

The preceding two studies focused on heart rate. The next four studies all included assessment of systolic and diastolic blood pressure responses in addition to heart rate response. Carmelli, Chesney, Ward, and Rosenman (1985) employed 12 MZ and 19 DZ male pairs recruited from the National Academy of Sciences–National Research Council (NAS–NRC) Twin Registry, each of whom completed a mental arithmetic task. On the basis of published mean squares, heritability estimates in the range of 20–60% can be obtained for both systolic and diastolic blood pressure responses. In contrast to the Carroll et al. (1985) and Turner et al. (1986) studies, the results of this study were consistent with zero heritability for heart rate reactivity. However, a later study by the same group (Carmelli et al., 1991), in which a considerably larger sample size was employed, resulted in a heart rate response heritability of 0.58 (mental arithmetic again being the stressor). The heritability estimates for systolic and diastolic blood pressure response were 0.80 and 0.66, respectively.

Smith et al. (1987) also employed a mental arithmetic stressor, which was completed by 82 MZ and 88 DZ male twin pairs. Calculation of intraclass correlations and subsequent calculation of heritability estimates led to reported values of 0.48 and 0.52 for systolic and diastolic blood pressure responses, respectively. For heart rate responses, no heritability estimate was calculated, since correlational analyses did not suggest a genetic influence.

The final study in this section is particularly noteworthy for two reasons that will be expanded on later in the chapter. For now, we shall simply note the results. Ditto (1993) employed 100 twin pairs. There were 20 MZ male, 20 DZ male, 20 MZ female, 20 DZ female, and 20 opposite-sex DZ pairs, and each completed a mental arithmetic task and a concept formation task. Maximum-likelihood path analytical techniques were used to reveal, in general, modest heritabilities for all three physiological response parameters to the stressors, although there was no evidence of familial correlation for systolic blood pressure responses to the concept formation task.

Discussion and Conclusions

Following this brief examination of each study (a more comprehensive analysis of these studies can be found in Turner and Hewitt, 1992), three overarching comments are appropriate. First, the ten studies reviewed are characterized by a wide array of subject samples. It is not always clear how particular subject samples were ascertained, and whether specific populations were targeted and then sampled or whether sampling was largely a matter of convenience. A notable exception, however, is the targeting of males in most of these studies. Until relatively recently, a bias toward the use of males in cardiovascular psychophysiological studies was also evident in nontwin research. Fortunately, this state of affairs is now changing. For comprehensive discussions of the literature on gender comparison, the reader is directed to recent reviews by Saab (1989) and Stoney (1992); here, we will concern ourselves specifically with the use of female subjects in twin studies.

As discussed in Chapter 10, the inclusion of both male and female twin pairs facilitates consideration of sex limitation of genetic or environmental influences or both. The Ditto (1993) study, mentioned above, is noteworthy because it permits the possible evaluation of sex limitation. Ditto's sample included not only same-sex male and female DZ twin pairs (MZ male and female pairs are, of course, ipso facto same sex), but also opposite-sex DZ pairs. Given the increasing interest in gender effects on both cardiovascular and neuroendocrine stress responses (Girdler, Turner, Sherwood, & Light, 1990; Light, Turner, Hinderliter, & Sherwood, 1993; van Doornen, 1986) and the appropriate statistical tools with which to examine a series of hypotheses about sex limitation of genetic and environmental influences [see Heath, Neale, Hewitt, Eaves, & Fulker (1989) and Chapter 10], it is to be hoped that more published reports of such studies will appear in the near future.

Second, the age composition of subject samples in the reviewed studies varies widely; it ranges from high school students (Vandenberg et al., 1965) to subjects around 60 years of age (Carmelli et al., 1985, 1991). In some studies (e.g., Carroll et al., 1985), young subjects (aged 16–24 years) were chosen, since cardiovascular disease, and hence secondary influences on cardiovascular functioning during stress, is minimally present in this age group. In other studies, subjects were chosen for different reasons. For example, Carmelli et al. (1985) chose subjects who had been extensively evaluated during a previous study. In yet other instances, no clear rationale is given for the age of chosen subjects. However, attention does need to be paid to this issue. Nongenetic psychophysiological studies have revealed that reactivity can vary with age (see Light, 1989). As for genetic studies, it is highly likely that there will be genotype–age interactions, and therefore careful consideration should be given to the sample's age composition. Particular attention should be directed toward young adults and middle-aged individuals, since there is now evidence of different genetic influences on anthropometric variables such as the body mass index between these age groups (Fabsitz, Carmelli, & Hewitt, 1992) (see also Chapter 8), and there may well be such differences for cardiovascular measures. Snieder and colleagues examine this topic in detail in Chapter Six.

Third, the experimental methods, and particularly the form of statistical analysis chosen, varied considerably among studies. Methods of analysis ranged from the calculation of correlations for each zygosity and the comparison of the resulting coefficients, through within-pair F-ratio variance comparisons of MZ and DZ twins and weighted least-squares model-fitting, to maximum-likelihood path analytical techniques. It is fair to note here that the analytical strategies offered by the discipline of behavior genetics are continuously evolving, and therefore the more recent studies have potential access to more recent statistical developments. Nevertheless, the study reported by Ditto (1993) should again serve as an exemplar here, since he took full advantage of recent advances in analytical approaches. Path analysis has become the technique of choice in this area, since it facilitates the study of sex effects, social interactions, between-group heterogeneity, and, of special importance to research in behavioral medicine, generalization from univariate to multivariate genetic analysis (see Heath et al., 1989; Hewitt, Stunkard, Carroll, Sims, & Turner, 1991; Neale & Cardon, 1992).

Bearing in mind all these considerations, what conclusions can be drawn on the basis of this body of literature? With due caution and attention to experimental procedures and sample characteristics, it may be concluded that (1) heart rate and blood pressure responses to laboratory psychological stressors are moderately heritable and (2) it is highly improbable that nongenetic influences shared by siblings play any substantial or lasting role in determining individual differences in stress responses. Examination of these studies also reveals two future research needs. First, since it is quite possible that there is sex limitation in the genetic and environmental determinants of cardiovascular stress responses, the design of future studies needs to take this possibility into account (see Chapter 10). Second, greater attention needs to be paid to the age of the subjects under study; it is probable that there are genotype–age interactions in the control of stress responses that are only just now beginning to be systematically explored and documented (see Chapter 6).

FUTURE DIRECTIONS

In this section of the chapter, we outline several strategies for future research. Since it is important to study cardiovascular activity both in the psychophysiological laboratory and in people's everyday surroundings, the suggested approaches informatively employ both laboratory investigation and ambulatory monitoring.

Laboratory Studies

The studies reviewed in the previous section of this chapter focused on heart rate and blood pressure. While these parameters were the traditional cardiovascular measures for many years, recent advances in the noninvasive measurement techniques of cardiovascular psychophysiology have provided the opportunity to study additional parameters. For example, much attention in current cardiovascular behavioral medicine is being directed toward the underlying hemodynamic determinants of blood pressure, namely, cardiac output and total peripheral resistance of the systemic vasculature.

Blood pressure is (figuratively and mathematically) the product of these two variables. Accordingly, a given increase in blood pressure (e.g., a given stress response) can be the result of an increase in cardiac output, an increase in total peripheral resistance, or a combination of alterations in both parameters. We want to know how a given individual (or group of individuals) typically arrives at a blood pressure increase. Two possibilities of interest are via the "myocardial" route (an increase in cardiac output) or via the "vascular" route (an increase in total peripheral resistance). These differing means of producing the same blood pressure alteration may have differing pathophysiological implications and consequences.

The technique of impedance cardiography (see Hurwitz et al., 1993; Sherwood et al., 1990; Wilson, Lovallo, & Pincomb, 1989) permits the noninvasive assessment of cardiac output and, in conjunction with contemporaneous blood pressure assessment, the derivation of total peripheral resistance. It also allows the measurement of systolic time intervals, an index of myocardial contractility. This technique is particularly well suited to behavioral medicine research (Sherwood, 1993), and its use in genetic studies is advocated (see also the discussion in Chapter Six).

Another possible improvement in future research concerns the choice of stressor tasks. While cognitive and visuomotor tasks (such as mental arithmetic and video games) have proved extremely informative, the latest generation of behavioral medicine tasks often incorporates psychosocial aspects. Tasks such as simulated public speaking, persuasion tasks, and current events discussion (Saab, Matthews, Stoney, & McDonald, 1989; Smith & Allred, 1989; Smith, Allred, Morrison, & Carlson, 1989) may allow the concurrent assessment of etiologically important personality characteristics and physiological parameters in situations that are similar to those arising naturally (Turner & Hewitt, 1992). Once again, a wider sampling of tasks will improve the informativeness of genetic studies of stress responses.

As a final example here, the incorporation of ethnic comparisons in twin studies may well prove informative. Recent observations suggest that African Americans (particularly African American males) show larger total peripheral resistance responses during laboratory stressors than their Caucasian American counterparts (Anderson, McNeilly, & Myers, 1992; Light et al., 1993). Accordingly, exploration of the inheritance of hemodynamic responses in both ethnic groups may prove helpful. It must be emphasized, however, that while attention to date has focused on comparison of African and Caucasian Americans, other ethnic groups should certainly be studied.

Laboratory–Field Generalization Studies

Laboratory–field generalization is concerned with whether the responses an individual displays to laboratory stressors are representative of those shown in everyday situations (for recent reviews of this field, see Turner, et al., 1994; van Doornen & Turner, 1992; Van Egeren & Gellman, 1992). To date, the degree of association between laboratory and real-world responses has typically been evaluated by correlational or multiple-regression analysis (see Turner et al., 1994). However, use of genetically informative samples and analyses allows a more sophisticated evaluation. Path analysis, which allows the generalization from univariate to multivariate genetic analysis, is the technique of choice here, since, although laboratory data can be analyzed in a univariate manner, the joint analysis of laboratory and field data requires a bivariate or multivariate approach.

Consider the example presented by Turner and Hewitt (1992). The classic twin design study permits the resolution of additive genetic effects (A), environmental influences shared by members of a family (C), and environmental influences unique to the individual (E). Figure 1 represents the linear path model for a phenotype, e.g., the systolic blood pressure response to a laboratory stressor, for a pair of twins P_1 and P_2. This model allows us to derive the expectations for the observed variances and covariances for MZ and DZ twins. The model can be fit to the data using a path-modeling program such as LISREL VII (Heath et al., 1989; Jöreskog & Sörbom, 1988) or Mx (Neale, 1993). This procedure will provide estimates of the partial regressions (*h, c,* and *e*) of the observed phenotype (P) on the unobserved genetic (A), shared environmental (C), and individual environment (E) variables, respectively. These in turn permit the calculation of the heritability estimate as $h^2/(h^2 + c^2 + e^2)$. Likelihood-ratio χ^2 tests of the

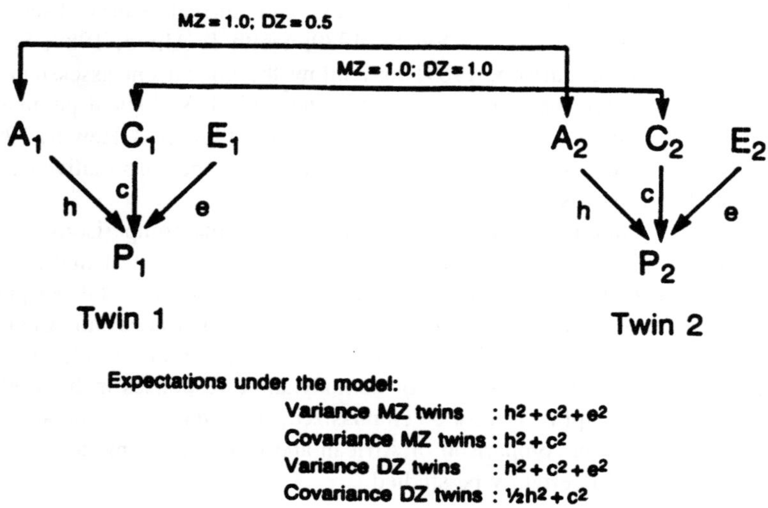

FIGURE 1. Univariate path model for the classic twin study. Reprinted by permission from J. Rick Turner and John K. Hewitt, "Twin Studies of Cardiovascular Response to Psychological Challenge: A Review and Suggested Future Directions," *Annals of Behavioral Medicine*, vol. 14, pp. 12–20. Copyright 1992 Society of Behavioral Medicine.

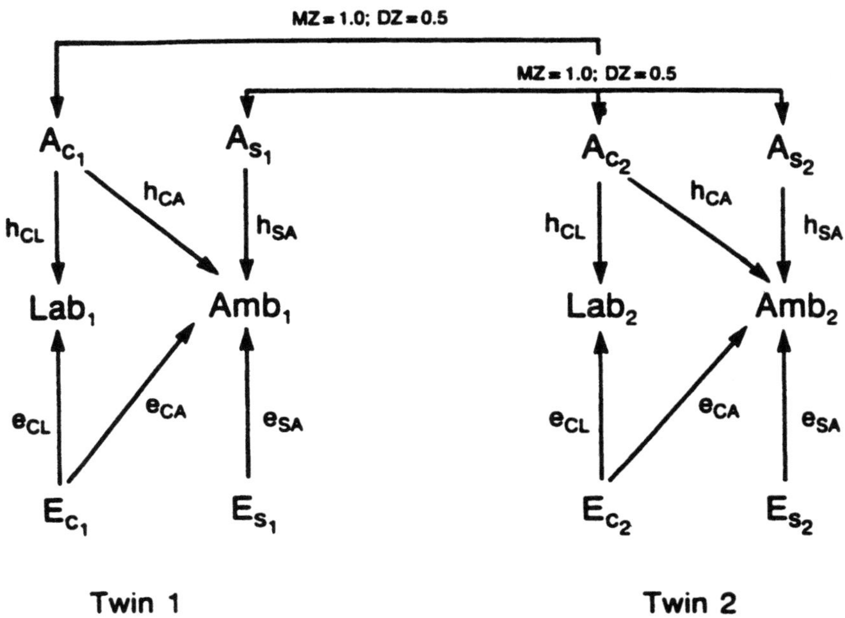

FIGURE 2. An illustrative bivariate path model for the classic twin study analysis of laboratory and ambulatory measures. Reprinted by permission from Turner and Hewitt (1992). Copyright 1992 Society of Behavioral Medicine.

adequacy of the model and of the significance of individual parameters are also provided along with standard errors of estimation.

The second step here concerns the use of multivariate approaches to consider laboratory and ambulatory data jointly. Figure 2 illustrates a bivariate genetic and environmental model for laboratory (Lab) and ambulatory (Amb) blood pressure. For the sake of simplicity, shared family environmental influences are not shown on this occasion. In the model are genetic influences (A_c) on the laboratory phenotype that have a common influence on the ambulatory phenotype. However, there are also new genetic influences (A_s) that are specific to the ambulatory phenotype. A similar decomposition is also described for the environmental influences (E_c and E_s). This model can be fit to the observed variance–covariance matrices for the MZ and DZ groups, again using maximum-likelihood model-fitting implemented by LISREL VII or Mx.

This analysis permits estimation of the univariate heritabilities along with estimates of the genetic and environmental covariates and correlations between the laboratory and ambulatory phenotypes. Of particular interest in the present context is that such an analysis provides a test for the presence of *additional* genetic or environmental influences, or both, on ambulatory blood pressure, over and above those influences on the laboratory blood pressure. Given the occurrence of these specific influences, *their* magnitudes can be estimated (along with the standard errors of estimation) and their relationship to (genetic *and* environmental) risk factors and outcomes evaluated.

AMBULATORY MONITORING STUDIES

Pickering (1989, 1991) and others have documented the value of ambulatory blood pressure data in cardiovascular behavioral medicine. Ambulatory monitoring (see Harshfield & Pulliam, 1992; Mancia, 1990; Pickering, 1992) is a topic of investigation in its own right as well as a technique that facilitates the investigation of laboratory–field generalization. Ambulatory monitoring provides the opportunity to study both *average levels* during normal daily activities and *variation* across time, situations, and differing physical and—of particular interest here—psychological states.

The richness of 24-hour ambulatory records allows a full investigation of blood pressure during the naturally occurring daily event cycle, which includes work and rest, activity and inactivity, wakefulness and sleep, and stress and quiescence. Laboratory investigation has demonstrated that an acute stressor can severely disrupt quiescent cardiovascular activity. It is unlikely, however, that one acute stressor will have a discernible influence on the decades-long progression of hypertensive disease [This logic certainly does not hold for other cardiovascular events: Powerful, single events can indeed precipitate myocardial infarctions (see Kamarck and Jennings, 1991).] As Pickering (1989, p. 270) observed:

> Although it has been shown that emotional changes can acutely alter blood pressure, the big question in behavioral hypertension research is whether or not psychological factors can contribute to sustained elevations of pressure.

The potential value of ambulatory monitoring is further supported by observations that target-organ damage is more closely related to ambulatory pressure than to clinic pressures (see Pickering & Devereux, 1987) and that ambulatory pressures may be better predictors of morbid events than clinic pressures (Perloff, Sokolow, & Cowan, 1983; Mann, Millar-Craig, & Raftery, 1985).

Although ambulatory monitoring seems to hold considerable promise as a method for studying responses to stress in everyday life, we are aware of only three published twin studies of ambulatory pressures. Rose (1984) reported a study of young adult twins who self-measured their casual blood pressure 6 times daily for 3–4 weeks. These twins also underwent ambulatory blood pressure monitoring whenever possible, primarily as a validity check of the self-measures. Although, as Rose (1984, p. 171) commented, "ambulatory records from normotensive twins have value in and of themselves," this study did not permit substantive conclusions to be drawn.

In the second study, Degaute, Van Cauter, van de Borne, and Linkowski (1994) reported observations on a small sample of young male twin pairs (28 MZ, 16 DZ) monitored while the twins were subjected to a standardized physical and social environment. Unfortunately for our purposes, however, the small sample sizes mean that any individual correlation cannot be taken seriously: For the 16 DZ pairs, the standard error of a correlation would be approximately 0.3. The fact that readers are asked to consider correlations for some 33 different measures for this sample makes interpretation more difficult, as does the possibility of undetermined biases in the sample indicated by the unusual finding of significantly higher blood pressures in the MZ than in the DZ twins. With these caveats in mind, the study's conclusion that there were genetic influences on some characteristics of the 24-hour profiles for diastolic blood pressures and heart rates again suggests that we need a more definitive genetic study of ambulatory blood pressure.

Finally, Somes, Harshfield, Alpert, Goble, and Schieken (1995) reported an analysis of ambulatory blood pressure data recorded every 20 minutes over a 24-hour period in 38 MZ and 28 DZ adolescent twin pairs. Various measures of intrapair differences were computed, and they showed greater MZ than DZ similarity, as would be expected if there were a significant genetic influence. Once again, this relatively small sample would not be expected to generate precise estimates of heritability, although it should be noted that on the basis of a comparison of same-sex DZ pairs and MZ pairs, heritability estimates for 24-hour means and for a measure of hour-to-hour variability ranged from a low of 0.08 to a high of 0.80, with an average around 0.3, for systolic pressure, diastolic pressure, and heart rate.

The paucity of adequate genetic studies on ambulatory measures no doubt reflects the relatively recent development of reliable monitoring technology designed to be used unobtrusively and comfortably by the subjects being monitored. Now that such equipment has become widely available, we can anticipate an increasing number of genetic studies that incorporate 24-hour ambulatory observations. The aims of such a study might include the investigation of the extent to which ambulatory blood pressure levels are genetically controlled and the extent to which they are environmentally conditioned. On one hand, we need to pay attention to potentially important short-term changes, such as the extent of early to midmorning rises in pressures, nocturnal decline in pressures, and other short-term responses that accompany different activity levels (Anastasiades & Johnston, 1990; Patterson et al., 1993) and postural changes (Sherwood & Turner, 1993; Turner & Sherwood, 1991) as well as psychological stresses. On the other hand, we need to consider longer-term influences on minimum and maximum diurnal pressures and average levels associated with anthropometric variables (Hinderliter, Willis, & Light, 1992; Gerber, Schnall, & Pickering, 1990) and life-style or personality variables.

Although such a study might concentrate on younger adults, we can assess the potential clinical significance of both stable individual differences in ambulatory pressures and short-term responsiveness through their relationship to family history for hypertension and coronary heart disease. Of course, such associations will be indicative of *genetic* correlations, or at least *familial* correlations, between early risk indices and later disease and will not identify risks of a purely non-familial environmental etiology. Despite this potential shortcoming, it seems improbable that a factor would be a risk for a disease outcome *only* when it did not aggregate in families; thus, the chances of missing an important potential risk factor altogether are likely to be small.

Beyond establishing the basic genetic and environmental patterns of variation, there are three further issues we might address: (1) the temporal stability of mean ambulatory blood pressures and their variability, (2) gender or sex limitation, and (3) ethnic group differences.

Repeated Ambulatory Measures

Two concerns are of relevance here. The first relates to the reproducibility of ambulatory blood pressure measurements. The temporal stability of *mean* ambulatory pressures has been reported to be quite high for periods of 4 months or less: Correlations across repeated sessions are reported to range from 0.54 to 0.80 for systolic pressure and from 0.36 to 0.93 for diastolic pressure (for a review, see Ward, Turner, & Johnston, 1994). The reproducibility of blood pressure *variability* appears to be less impressive.

The second concern relates to the degree to which there are true long-term changes in ambulatory mean pressures and their variability (as opposed to measurement error or day-to-day fluctuations). In order to distinguish change from measurement error and day-to-day fluctuations, several assessments (at least 3 and preferably 4 or more) (Hewitt, Eaves, Neale, & Meyer, 1988) are needed (i.e., several days' monitoring). To observe true change over a meaningful period, a period of at least 1 year seems necessary. A period of 12 months or more may also allow control for seasonal or other annual fluctuations. Hence, monitoring subjects' ambulatory pressures on 3 or more days a year might be a reasonable strategy.

Gender and Sex Limitation

A primary interest in nongenetic studies of gender differences concerns mean differences in blood pressure. While this is certainly of interest in genetic studies, we would like to go further and examine two other salient possibilities: (1) that different genetic or environmental factors act to cause variation in men than act in women and (2) that the magnitude of the impact of genetic and environmental variation differs in men and women. The importance of these questions and the research strategy that behavior genetics offers to address them are taken up in Chapter Ten.

Ethnicity

In the laboratory, all subjects are (quite deliberately) presented with exactly the same stressor(s). It is true that the newer generation of behavioral medicine tasks often contain the element of "standardized flexibility" (Turner, 1994), which allows each individual subject to receive one (possibly unique) permutation of the task depending on the individual's level of task-relevant skills. Nevertheless, in a very real sense, all subjects are presented with an environment that is standard. This is clearly not the case in ambulatory studies. When asked by the experimenter to go about their usual daily business, subjects' experiences will differ widely. Certainly, ethnicity is by no means the only determining factor here, but it may be associated with some of the differences during the course of the day. Anderson et al. (1992, p. 131) have suggested that in behavioral medicine research, ethnicity "should be viewed as a proxy for the effects of differential exposure to chronic social and environmental stressors rather than as a proxy for the effects of genetic differences." As these authors noted, African Americans, on average, are exposed to a wider array of chronic stressors than their Caucasian American counterparts.

These considerations are particularly pertinent to ambulatory studies, when data are collected during people's everyday activities in their habitual surroundings. For these reasons, we might anticipate that the relative contributions of genetic and environmental variation to individual differences in blood pressure, and their associated risk factors and disease outcomes, might differ between ethnic groups (see Chapter Eleven).

SUMMARY

Behavior genetic techniques can be used to great advantage in cardiovascular behavioral medicine. Laboratory-based twin studies have revealed that heart rate and blood pressure responses to psychological stressors show moderate heritability. Similar studies employing a wider range of physiological parameters may prove very informative, as may studies that systematically examine the influence of age.

Ambulatory monitoring of cardiovascular activity is equally important in contemporary stress research, yet very few genetic studies employing this approach have been reported to date. Accordingly, we have made some suggestions for future studies. Ambulatory data collected using the appropriate genetically informative designs may also cast important light on issues of both gender and ethnic group differences in the origins of risks for hypertension and coronary heart disease. Combining the measurement techniques of cardiovascular psychophysiology, the experimental paradigms of behavioral medicine, and the analytical techniques of behavioral genetics appears to be a particularly promising research strategy for investigating psychosomatic hypotheses of cardiovascular disease.

REFERENCES

Anastasiades, P., & Johnston, D. W. (1990). A simple activity measure for use with ambulatory subjects. *Psychophysiology, 27*, 87–93.

Anderson, N. B., McNeilly, M., & Myers, H. (1992). Toward understanding race differences in autonomic reactivity: A proposed contextual model. In J. R. Turner, A. Sherwood, & K. C. Light (Eds.), *Individual differences in cardiovascular response to stress* (pp. 125–145). New York: Plenum Press.

Blascovich, J., & Katkin, E. S. (1993). *Cardiovascular reactivity to psychological stress and disease*. Washington, DC: American Psychological Association.

Boomsma, D. I. (1992). *Quantitative genetic analysis of cardiovascular risk factors in twins and their parents*. Thesis: Free University of Amsterdam. Enschede: Febodruk.

Carmelli, D., Chesney, M. A., Ward, M. M., & Rosenman, R. H. (1985). Twin similarity in cardiovascular stress response. *Health Psychology, 4*, 413–423.

Carmelli, D., Ward, M. M., Reed, T., Grim, C. E., Harshfield, G. A., & Fabsitz, R. R. (1991). Genetic effects on cardiovascular responses to cold and mental activity in late adulthood. *American Journal of Hypertension, 4*, 239–244.

Carroll, D., Hewitt, J. K., Last, K. A., Turner, J. R., & Sims, J. (1985). A twin study of cardiac reactivity and its relationship to parental blood pressure. *Physiology and Behavior, 34*, 103–106.

Degaute, J.-P., Van Cauter, E., van de Borne, P., & Linkowski, P. (1994). Twenty-four-hour blood pressure and heart rate profiles in humans: A twin study. *Hypertension, 23*, 244–253.

Ditto, B. (1993). Familial influences on heart rate, blood pressure, and self-report anxiety responses to stress: Results from 100 twin pairs. *Psychophysiology, 30*, 635–645.

Eaves, L. J., Last, K. A., Young, P. A., & Martin, N. G. (1978). Model-fitting approaches to the analysis of human behaviour. *Heredity, 41*, 249–320.

Fabsitz, R. R., Carmelli, D., & Hewitt, J. K. (1992). Evidence for independent genetic influences on obesity in middle age. *International Journal of Obesity, 16*, 657–666.

Gerber, I. M., Schnall, P. L., & Pickering, T. G. (1990). Body fat and its distribution in relation to casual and ambulatory blood pressure. *Hypertension, 15*, 508–513.

Girdler, S. S., Turner, J. R., Sherwood, A., & Light, K. C. (1990). Gender differences in blood pressure control during a variety of behavioral stressors. *Psychosomatic Medicine, 52*, 571–591.

Harshfield, G. A., & Pulliam, D. A. (1992). Individual differences in ambulatory blood pressure patterns. In J. R. Turner, A. Sherwood, & K. C. Light (Eds.), *Individual differences in cardiovascular response to stress* (pp. 51–61). New York: Plenum Press.

Heath, A. C., Neale, M. C., Hewitt, J. K., Eaves, L. J., & Fulker, D. W. (1989). Testing structural equation models for twin data using LISREL. *Behavior Genetics, 19*, 9–35.

Hewitt, J. K., Eaves, L. J., Neale, M. C., & Meyer, J. M. (1988). Resolving causes of developmental continuity or "tracking." I. Longitudinal twin studies during growth. *Behavior Genetics, 18*, 122–151.

Hewitt, J. K., Stunkard, A. J., Carroll, D., Sims, J., & Turner, J. R. (1991). A twin study approach towards understanding genetic contributions to body size and metabolic rate. *Acta Geneticae Medicae et Gemellalogiae, 40*, 133–146.

Hinderliter, A. L., Willis, P. W., & Light, K. C. (1992). Relationships of obesity and blood pressure to left ventricular function in young subjects with normal or marginally elevated blood pressure. *American Journal of Hypertension, 5*, 31A.

Hines, E. A., Jr., McIlhaney, M. L., & Gage, R. P. (1957). A study of twins with normal blood pressures and with hypertension. *Transactions of the Association of American Physicians, 70*, 282.

Hurwitz, B. E., Nelesen, R. A., Saab, P. G., Nagel, J. H., Spitzer, S. B., Gellman, M. D., McCabe, P. M., Phillips, D. J., & Schneiderman, N. (1993). Differential patterns of dynamic cardiovascular regulation as a function of task. *Biological Psychology, 36*, 75–95.

Jöreskog, K. G., & Sörbom, D. (1988). *LISREL VII: A guide to the program and applications*. Chicago: SPSS Inc.

Kamarck, T. W., & Jennings, J. R. (1991). Biobehavioral factors in sudden cardiac death. *Psychological Bulletin, 109*, 42–75.

Light, K. C. (1989). Constitutional factors relating to differences in cardiovascular response. In

N. Schneiderman, S. M. Weiss, & P. G. Kaufmann (Eds.), *Handbook of research methods in cardiovascular behavioral medicine* (pp. 417–431). New York: Plenum Press.

Light, K. C., Turner, J. R., Hinderliter, A., & Sherwood, A. (1993). Race and gender comparisons: I. Hemodynamic responses to a series of stressors. *Health Psychology, 12*, 354–365.

Lovallo, W. R. (1975). The cold pressor test and autonomic function: A review and integration. *Psychophysiology, 12*, 268–282.

Mancia, G. (1990). Ambulatory blood pressure monitoring: Research and clinical applications. *Journal of Hypertension, 8*(Supplement 7), S1–S13.

Mann, S., Millar-Craig, M. W., & Raftery, E. B. (1985). Superiority of 24-hour measurement of blood pressure over clinical values in determining prognosis in hypertension. *Clinical and Experimental Hypertension, A7(2–3)*, 279–281.

Manuck, S. B. (1994). Cardiovascular reactivity in cardiovascular disease: "Once more unto the breach." *International Journal of Behavioral Medicine, 1*, 4–31.

Manuck, S. B., Kasprowicz, A. L., Monroe, S. M., Larkin, K. T., & Kaplan, J. R. (1989). Psychophysiological reactivity as a dimension of individual differences. In N. Schneiderman, S. M. Weiss, & P. G. Kaufmann (Eds.), *Handbook of research methods in cardiovascular behavioral medicine* (pp. 365–382). New York: Plenum Press.

Matthews, K. A., Rakaczky, C. J., Stoney, C. M., & Manuck, S. B. (1987). Are cardiovascular responses to behavioral stressors a stable individual difference variable in childhood? *Psychophysiology, 24*, 464–473.

Matthews, K. A., Weiss, S. M., Detre, T., Dembroski, T. M., Falkner, B., Manuck, S. B., & William, R. B., Jr. (1986). *Handbook of stress, reactivity, and cardiovascular disease*. New York: John Wiley.

McIlhaney, M. L., Shaffer, J. W., & Hines, E. A., Jr. (1975). The heritability of blood pressure: An investigation of 200 pairs of twins using the cold pressor test. *Johns Hopkins Medical Journal, 136*, 57–64.

Neale, M. C. (1993). *Mx: Statistical modelling*. Richmond: Virginia Commonwealth University, Departments of Human Genetics and Psychiatry.

Neale, M. C., & Cardon, L. R. (1992). *Methodology for genetic studies of twins and families*. Dordrecht, The Netherlands: Kluwer Academic Publishers.

Obrist, P. A. (1981). *Cardiovascular psychophysiology: A perspective*. New York: Plenum Press.

Patterson, S. M., Krantz, D. S., Montgomery, L. C., Deuster, P. A., Hedges, S. M., & Nebel, L. E. (1993). Automated physical activity monitoring: Validation and comparison with physiological and self-reported measures. *Psychophysiology, 30*, 296–305.

Perloff, D. Sokolow, M., & Cowan, R. (1983). The prognostic value of ambulatory blood pressures. *Journal of the American Medical Association, 249*, 2793–2798.

Pickering, T. G. (1989). Ambulatory monitoring: Applications and limitations. In N. Schneiderman, S. M. Weiss, & P. G. Kaufmann (Eds.), *Handbook of research methods in cardiovascular behavioral medicine* (pp. 261–272). New York: Plenum Press.

Pickering, T. G. (1991). *Ambulatory monitoring and blood pressure variability*. London: Science Press.

Pickering, T. G. (1992). The ninth Sir George Pickering memorial lecture: Ambulatory monitoring and the definition of hypertension. *Journal of Hypertension, 10*, 401–409.

Pickering, T. G., & Devereux, R. (1987). Ambulatory monitoring of blood pressure as a predictor of cardiovascular risk. *American Heart Journal, 114*, 925–928.

Rose, R. J. (1984). Familial influence on ambulatory blood pressure: Studies of normotensive twins. In M. A. Weber & J. I. M. Drayer (Eds.), *Ambulatory blood pressure monitoring* (pp. 167–172). New York: Springer-Verlag.

Rose, R. J. (1986). Familial influences on cardiovascular reactivity. In K. A. Matthews, S. M. Weiss, T. Detre, T. M. Dembroski, B. Falkner, S. B. Manuck, & R. B. Williams, Jr. (Eds.), *Handbook of stress, reactivity, and cardiovascular disease* (pp. 259–272). New York: John Wiley.

Rose, R. J., Grim, C. E., & Miller, J. Z. (1984). Familial influences on cardiovascular stress reactivity: Studies of normotensive twins. *Behavioral Medicine Update, 6,* 21–24.

Saab, P. G. (1989). Cardiovascular and neuroendocrine responses to challenge in males and females. In N. Schneiderman, S. M. Weiss, & P. G. Kaufmann (Eds.), *Handbook of research methods in cardiovascular behavioral medicine* (pp. 453–481). New York: Plenum Press.

Saab, P. G., Llabre, M. M., Hurwitz, B. E., Schneiderman, N., Wohlgemuth, W., Durel, L. A., Massie, C., & Nagel, J. (1993). The cold pressor test: Vascular and myocardial response patterns and their stability. *Psychophysiology, 30,* 366–373.

Saab, P. G., Matthews, K. A., Stoney, C. M., & McDonald, R. H. (1989). Premenopausal and postmenopausal women differ in their cardiovascular and neuroendocrine responses to behavioral stressors. *Psychophysiology, 26,* 270–280.

Shapiro, A. P., Nicotero, J., & Scheib, E. T. (1968). Analysis of the variability of blood pressure, pulse rate, and catecholamine responsivity in identical and fraternal twins. *Psychosomatic Medicine, 30,* 506–520.

Sherwood, A. (1993). The use of impedance cardiography in cardiovascular reactivity research. In J. Blascovich & E. S. Katkin (Eds.), *Cardiovascular reactivity to psychological stress and disease* (pp. 157–199). Washington, DC: American Psychological Association.

Sherwood, A., Allen, M. T., Fahrenberg, J., Kelsey, R. M., Lovallo, W. R., & van Doornen, L. J. P. (1990). Committee report: Methodological guidelines for impedance cardiography. *Psychophysiology, 27,* 1–23.

Sherwood, A., & Turner, J. R. (1992). A conceptual and methodological overview of cardiovascular reactivity research. In J. R. Turner, A. Sherwood, & K. C. Light (Eds.), *Individual differences in cardiovascular response to stress* (pp. 3–32). New York: Plenum Press.

Sherwood, A., & Turner, J. R. (1993). Postural stability of hemodynamic responses during mental challenge. *Psychophysiology, 30,* 237–244.

Smith, T. W., & Allred, K. D. (1989). Blood-pressure responses during social interaction in high- and low-cynically hostile males. *Journal of Behavioral Medicine, 12,* 135–143.

Smith, T. W., Allred, K. D., Morrison, C. A., & Carlson, S. D. (1989). Cardiovascular reactivity and interpersonal influence: Active coping in a social context. *Journal of Personality and Social Psychology, 56,* 209–218.

Smith, T. W., Turner, C. W., Ford, M. H., Hunt, S. C., Barlow, G. K., Stults, B. M., & Williams, R. R. (1987). Blood pressure reactivity in adult male twins. *Health Psychology, 6,* 209–220.

Somes, G. W., Harshfield, G. A., Alpert, B. S., Goble, M. M., & Schieken, R. M. (1995). Genetic influences on ambulatory blood pressure patterns: The Medical College of Virginia twin study. *American Journal of Hypertension* (in press).

Stoney, C. M. (1992). The role of reproductive hormones in cardiovascular and neuroendocrine function during behavioral stress. In J. R. Turner, A. Sherwood, & K. C. Light (Eds.), *Individual differences in cardiovascular response to stress* (pp. 147–163). New York: Plenum Press.

Theorell, T., deFaire, U., Schalling, D., Adamson, U., & Askevold, F. (1979). Personality traits and psychophysiological reactions to a stressful interview in twins with varying degrees of coronary heart disease. *Journal of Psychosomatic Research, 23,* 89–99.

Turner, J. R. (1989). Individual differences in heart rate response during behavioral challenge. *Psychophysiology, 26,* 497–505.

Turner, J. R. (1994). *Cardiovascular reactivity and stress: Patterns of physiological response.* New York: Plenum Press.

Turner, J. R., Carroll, D., Sims, J., Hewitt, J. K., & Kelly, K. A. (1986). Temporal and inter-task consistency of heart rate reactivity during active psychological challenge: A twin study. *Physiology and Behavior, 38,* 641–644.

Turner, J. R., & Hewitt, J. K. (1992). Twin studies of cardiovascular response to psychological challenge: A review and suggested future directions. *Annals of Behavioral Medicine, 14,* 12–20.

Turner, J. R., & Sherwood, A. (1991). Postural effects on blood pressure reactivity: Implications for studies of laboratory–field generalization. *Journal of Psychosomatic Research, 35,* 289–295.

Turner, J. R., Sherwood, A., & Light, K. C. (Eds.) (1992). *Individual differences in cardiovascular response to stress*. New York: Plenum Press.

Turner, J. R., Ward, M. M., Gellman, M. D., Johnston, D. W., Light, K. C., & van Doornen, L. J. P. (1994). The relationship between laboratory and ambulatory cardiovascular activity: Current evidence and future directions: *Annals of Behavioral Medicine, 16,* 12–23.

Vandenberg, S. G., Clark, P. J., & Samuels, I. (1965). Psychophysiological reactions of twins: Hereditary factors in galvanic skin resistance, heartbeat, and breathing rates. *Eugenics Quarterly, 12,* 7–10.

van Doornen, L. J. P. (1986). Sex differences in physiological reactions to real life stress and their relationship to psychological variables. *Psychophysiology, 23,* 657–662.

van Doornen, L. J. P., & Turner, J. R. (1992). The ecological validity of laboratory stress testing. In J. R. Turner, A. Sherwood, & K. C. Light (Eds.), *Individual differences in cardiovascular response to stress* (pp. 63–83). New York: Plenum Press.

Van Egeren, L. F., & Gellman, M. D. (1992). Cardiovascular reactivity to everyday events. In E. H. Johnson, W. D. Gentry, & S. Julius (Eds.), *Personality, elevated blood pressure, and hypertension* (pp. 135–150). Washington, D.C.: Hemisphere Publishing Corp.

Ward, M. M., Turner, J. R., & Johnston, D. W. (1994). Temporal stability of ambulatory cardiovascular monitoring. *Annals of Behavioral Medicine, 16,* 3–11.

Wilson, M. F., Lovallo, W. R., & Pincomb, G. A. (1989). Noninvasive measurement of cardiac function. In N. Schneiderman, S. M. Weiss, & P. G. Kaufmann (Eds.), *Handbook of research methods in cardiovascular behavioral medicine* (pp. 23–50). New York: Plenum Press.

CHAPTER SIX

Developmental Genetic Trends in Blood Pressure Levels and Blood Pressure Reactivity to Stress

HAROLD SNIEDER, LORENZ J. P. VAN DOORNEN, AND DORRET I. BOOMSMA

INTRODUCTION

This chapter has two main aims. The first is to describe the changes in heritability of blood pressure levels and blood pressure reactivity that occur during the life span. The second is to disentangle the genetic and nongenetic causes of stability and change in these parameters. In order to achieve these goals, both empirical studies and developmental models will be discussed.

After an initial examination of blood pressure levels and reactivity, the results of twin studies on subjects of different ages are compared to address the question of whether heritabilities are age-specific. The next section focuses on parent–offspring studies and studies with a special interest in age-dependent genetic and environmental effects. Following this presentation, the very limited number of longitudinal genetic studies of blood pressure are discussed, and an extended parent–offspring design is described and illustrated with relevant data from our own laboratory. This design, coupled with the appropriate modeling, shows how one can estimate genetic stability in the absence of longitudinal data.

HAROLD SNIEDER, LORENZ J. P. VAN DOORNEN, AND DORRET I. BOOMSMA • Department of Psychophysiology, Free University of Amsterdam, De Boelelaan 1111, 1081 HV Amsterdam, The Netherlands.

Behavior Genetic Approaches in Behavioral Medicine, edited by J. Rick Turner, Lon R. Cardon, and John K. Hewitt. Plenum Press, New York, 1995.

Finally, we discuss the implications of such studies and speculate about what directions such research might take in the future.

BLOOD PRESSURE LEVELS

Blood pressure levels have been shown to be related to the risk of coronary heart disease and stroke. Risk increases linearly from a systolic blood pressure (SBP) of 110 mm Hg on, and even increases progressively in the higher pressure range (Pooling Project Research Group, 1978; Kannel, 1976). In Western societies, blood pressure rises with age. This trend, however, is not a simple linear one, and it differs between males and females. In boys, a strong rise in SBP of about 15 mm Hg occurs in the first 8 weeks of life; in girls, this rise occurs earlier and is even steeper. In both sexes, SBP then remains stable until the age of 12 months. Diastolic blood pressure (DBP) declines after birth in both boys and girls, reaching its minimum in 2–3 months, but is back at birth level by the end of the first year. From then on, there is a gradual increase in SBP (+20 mm Hg) and DBP (+15 mm Hg) in both sexes until puberty.

During adolescence, SBP again rises, this time more steeply in boys than in girls. At the age of 20, SBP in boys is nearly 10 mm Hg higher than in girls, whereas DBP is only slightly higher (Lauer, Burns, Clarke, & Mahoney, 1991; Labarthe, Mueller, & Eissa, 1991; van Lenthe & Kemper, 1993). These sex differences remain until the age of 50.

During the middle-age period (40–60 years), SBP rises about 10 mm Hg in males, and the variance in values also increases. DBP levels and their variance remain fairly constant in this period (Pooling Project Research Group, 1978). Recently, these age trends in levels and variances during middle age were confirmed in a large Norwegian sample and shown to apply to females as well (Tambs et al., 1992). The SBP rise from 40 years of age on is steeper in females. After the age of 50, women have a higher SBP than men. Menopause as such does not seem to be responsible for this difference (Lerner & Kannel, 1986; Matthews et al., 1989). SBP, but not DBP, continues to rise until the age of 70. After this age, both SBP and DBP decline. In later life, females have higher SBP and DBP than males (Valkenburg, Hofman, Klein, & Groustra, 1980).

The age-specific increase in SBP and DBP, and the sex differences in this increase, suggest that different mechanisms have their influence on blood pressure in different periods of life and that they may not be the same, or have the same timing, in males and females. The balance between genetic and environmental influences on blood pressure thus may also vary as a function of age and sex. There may even be "critical periods" in which exposure to certain environmental factors, or the expression of particular genes, will have a decisive impact on future blood pressure, whereas exposure to the same factors in other stages of life may be relatively inconsequential (Unger & Rettig, 1990).

BLOOD PRESSURE REACTIVITY

Exaggerated cardiovascular reactivity to stress is hypothesized to play some role in the development of hypertension (Manuck, Kasprowicz, & Muldoon, 1990). Support for this idea is furnished by studies showing that increased cardiovascular reactivity to stress is associated with stronger rises of blood pressure with age. This was shown by Menkes et al. (1989) for the response to the cold pressor test and by Matthews, Woodall, and Allen (1993) for the response to mental stressors. Promising as these results may be, these observations do not necessarily imply a causal role of exaggerated stress reactivity for the development of high blood pressure. Such stress reactivity could be just a risk marker reflecting underlying structural or functional vascular changes (Folkow, Hallback, Lundgren, & Weiss, 1970; Adams, Bobik, & Korner, 1989).

When investigating the contribution of genetic factors to blood pressure reactivity, it is again of interest to know whether this influence varies with age. A change in reactivity with age would perhaps point to this possibility. A decline in heart rate responsiveness to mental stress with increasing age has been observed in several studies (Faucheux et al., 1989; Garwood, Engel, & Capriotti, 1982; Barnes, Raskind, Gumbrecht, & Halter, 1982; Furchtgott & Busemeyer, 1979; Gintner, Hollandsworth, & Intrieri, 1986). This observation fits the general finding of attenuated cardiac responsiveness to β-receptor agonists. For blood pressure reactivity, one would predict an increased stress reactivity with age on physiological grounds. α_1-Receptor-mediated functions are preserved with aging, whereas β-receptor-mediated vasodilation and arterial compliance decrease with age (Folkow & Svanborg, 1993). The empirical support for an age trend in blood pressure reactivity is, however, not convincing. Matthews and Stoney (1988) found larger SBP responses in adults as compared to children, but no association within the adult age range (31–62). Garwood et al. (1982) observed a small positive correlation of SBP reactivity with age ($r = 0.25$) in the range between 30 and 79 years. Other studies, however, found no age effect (Faucheux et al., 1989; Gintner et al., 1986; Steptoe & Ross, 1981). In none of these studies was DBP reactivity age-dependent.

The absence of a convincing age trend in blood pressure reactivity does not refute, however, the possibility that genetic or environmental influences on reactivity, or both, differ by age. A role for genetic factors in stress reactivity finds support in studies showing that a family history of hypertension in subjects who are still normotensive themselves is associated with higher blood pressure reactivity to stressors (De Visser, 1994).

Currently, we know very little about developmental trends in the heritability of important cardiovascular and other (psycho)physiological parameters, and we know even less about causes of stability and change in these variables. As stated, the age trends in blood pressure level may indicate the involvement of different genetic or environmental factors, or both, at different ages. The same applies to the increases in phenotypic variance, especially for SBP, across the life span. This increase may be due to an increase in the amount of genetic variance, nongenetic variance, or both. Such changes in variance components can imply changes in heritabilities with age ($h^2 = Vg/Vp$, where h^2 = heritability, Vg = genetic variance, and Vp = total phenotypic variance) as well as differences in correlations among relatives of different ages. For example, if the amount of genetic variance is the same at different ages while the amount of environmental variance increases as people grow older, the covariance between parents and offspring will be the same as the covariance among siblings (assuming that the same genes are expressed at different ages and that shared environmental and dominance influences are negligible). However, the parent–offspring *correlation* will be lower than the sibling correlation, due to the increase in variance in the parental generation.

Changes in phenotypic variance and in the part of the variance that may be attributed to genetic and environmental factors, i.e., heritability and environmentality (Plomin, DeFries, & McClearn, 1990), may be detected in cross-sectional or longitudinal studies. However, only longitudinal studies, in which the same subjects are measured repeatedly, are informative about the stability of genetic and environmental factors. Contrary to popular points of view, genetically determined characteristics need not be stable, nor are longitudinally stable characteristics always influenced by heredity (Molenaar, Boomsma, & Dolan, 1991). During development, stability of a quantitative trait such as blood pressure may be due to stable environmental influences, while instability may be due to distinct subsets of genes turning on and off. This chapter considers the kinds of empirical evidence that will help us to understand the genetic and environmental influences on the stability and change of blood pressure as a dynamic characteristic.

TWIN STUDIES

Blood Pressure Levels

If twins within a specific age range are measured in studies estimating the genetic influence on blood pressure level, heritability values for this specific age range are obtained. The available twin studies are listed in Table 1. These studies did not take an explicit interest in age, but studies are listed in ascending order according to age of the twin sample.

Such a systematic overview of all studies may reveal any age-dependent trends in heritability. If studies considered sex differences in h^2, estimates for males and females are listed separately. The studies in Table 1 found no evidence for influence of shared family environment (c^2) on blood pressure. With respect to heritability estimates, the results are remarkably consistent: The majority of heritability estimates lie between 0.40 and 0.70 for both males and females, and no clear age trend in h^2 can be detected.

TABLE 1. Twin Studies Estimating Heritability in Systolic Blood Pressure and Diastolic Blood Pressure, in Ascending Order According to Age[a]

		Age			Heritability (h^2)	
Investigator	Pairs of twins	Mean (SD)	Range	Sex	SBP	DBP
Levine et al. (1982)	67 MZ, 99 DZ	? (?)	0.5–1.0	M & F	0.66	0.48
Schieken et al. (1989)	74 MZF, 71 MZM,	11.1 (0.25)	?	M	0.66	0.64
	31 DZF, 23 DZM, 52 DOS			F	0.66	0.51
McIlhany, Shaffer, & Hines (1975)	47 MZF, 40 MZM,	14.0 (6.5)	5.0–50.0	M	0.41	0.56
	36 DZF, 32 DZM, 45 DOS			F	0.78	0.61
Boomsma (see the text) (1992)	33 MZF, 35 MZM,	16.8 (2.0)	13.0–22.0	M	0.49	0.69
	29 DZF, 31 DZM, 28 DOS			F	0.66	0.50
Sims et al. (1987)	40 MZM, 45 DZM	19.4 (3.0)	?		0.68	0.76
Ditto (1993)	20 MZF, 20 MZM,	20.0 (5.0)	12.0–44.0	M	0.63	0.58
	20 DZF, 20 DZM, 20 DOS			F	0.63	0.58
Bielen et al. (1991)	32 MZM	21.7 (3.7)	18.0–31.0		0.69	0.32
	21 DZM	23.8 (3.9)				
Hunt et al. (1989)	73 MZM, 81 DZM	34.5 (9.5)	?		0.54	0.60
Slattery et al. (1988)	77 MZM, 88 DZM	? (?)	22.0–66.0		0.60	0.66
Snieder et al. (see the text)	47 MZF, 43 MZM,	44.4 (6.7)	34.0–63.0	M	0.40	0.42
	39 DZF, 32 DZM, 39 DOS			F	0.63	0.61
Feinleib et al. (1977)	250 MZM, 264 DZM	? (?)	42.0–56.0		0.60	0.61
Theorell, DeFaire, Schalling, Adamson, & Askevold (1979)	17 MZM, 13 DZM	62.0 (?)	51.0–74.0		0.00	0.00

[a] Abbreviations: (MZF) monozygotic females; (MZM) monozygotic males; (DZF) dizygotic females; (DZM) dizygotic males; (DOS) dizygotic opposite-sex; (SBP) systolic blood pressure; (DBP) diastolic blood pressure; (M) male; (F) female.

TABLE 2. Twin Studies Estimating Heritability in Systolic Blood Pressure and Diastolic Blood Pressure Reactivity, in Ascending Order According to Age[a]

		Age				Heritability (h^2)	
Investigator	Pairs of twins	Mean (SD)	Range	Task	Sex	SBP	DBP
McIlhany et al. (1975)	47 MZF, 40 MZM, 36 DZF, 32 DZM, 45 DOS	14.0 (6.5)	5.0–50.0	CP	M	0.36	0.53
					F	0.72	0.68
Boomsma (see the text) (1992)	33 MZF, 35 MZM, 29 DZF, 31 DZM, 28 DOS	16.8 (2.0)	13.0–22.0	RT	M	0.00	0.00
					F	0.00	0.00
				MA	M	0.44	0.38
					F	0.44	0.38
Ditto (1993)	20 MZF, 20 MZM, 20 DZF, 20 DZM, 20 DOS	20.0 (5.0)	12.0–44.0	MA	M	0.36	0.47
					F	0.36	0.47
				CT	M	0.00	0.34
					F	0.00	0.34
				IH	M	0.00	0.57
					F	0.00	0.57
				CP	M	0.38	0.81
					F	0.38	0.22
Shapiro et al. (1968)	7 MZF, 5 MZM	23.6 (?)	?	ST	M & F	0.70 (= MAP)	
	8 DZF, 4 DZM	21.9 (?)	?				
Smith et al. (1987)	82 MZM, 88 DZM	35.0 (?)	21.0–61.0	MA		0.48	0.52
Snieder et al. (see the text)	47 MZF, 43 MZM, 39 DZF, 32 DZM, 39 DOS	44.4 (6.7)	34.0–63.0	RT	M	0.37	0.23
					F	0.27	0.23
				MA	M	0.36	0.25
					F	0.36	0.25
Theorell et al. (1979)	17 MZM, 13 DZM	62.0 (?)	51.0–74.0	SI		0.00	0.00
Carmelli et al. (1991)	47 MZM, 54 DZM	62.4 (?)	59.0–69.0	MA		0.80	0.66

[a] Abbreviations: (CP) cold pressor task; (RT) reaction time task; (MA) mental arithmetic task; (CT) concept task; (IH) isometric handgrip; (ST) Stroop task; (SI) structured interview; (MAP) mean arterial pressure. See the Table 1 footnote for further abbreviations.

Some caution concerning the interpretation of Table 1 is required. First, studies used different methods to estimate h^2. Second, there was wide variation in the composition of the twin samples in the studies: The age range within studies differs considerably, and results are more reliable for males, since the majority of studies used only male twins. Furthermore, results from studies with small sample sizes are clearly less reliable. Nevertheless, the absence of an age trend in h^2 in Table 1 seems to warrant the conclusion that the relative influence of genes on blood pressure does not change appreciably with age.

Blood Pressure Reactivity

Studies that have investigated genetic influence on blood pressure reactivity are scarce. Turner and Hewitt (1992) (see also Chapter 5) reviewed the twin studies of blood pressure response to psychological stress. They concluded that:

1. Blood pressure reactivity to psychological stress is moderately heritable.
2. A common environmental influence is highly improbable.
3. Sex differences are possible, but as yet unexplored.
4. Genotype–age interactions are probable, but they await systematic evaluation.

A first impression of this genotype–age interaction can be obtained by listing the twin studies in ascending order according to the subjects' ages (Table 2).

Heritability estimates of blood pressure reactivity vary between 0.00 and 0.81 and are far less consistent than estimates for blood pressure levels. As in Table 1, a trend in h^2 for blood pressure reactivity with age cannot be readily detected. In addition to the interpretational difficulties mentioned in relation to Table 1, interpretation of Table 2 is hampered by the use of different stress tasks across studies. Even the use of different stress tasks within the same study leads to different h^2 estimates (Ditto, 1993).

PARENT–OFFSPRING STUDIES

Blood Pressure Levels

Another approach to investigating the age dependency of genetic and environmental effects is to compare parent–offspring data with data from siblings or twins (Eaves, Last, Young, & Martin, 1978). If there is an age-dependent genetic or environmental effect on the phenotype, one would expect the parent–offspring correlation to be lower than sibling or dizygotic (DZ) twin correlations, as the latter are measured around the same age. This expectation was confirmed in a review by Iselius, Morton, and Rao (1983). They pooled the results from a number of studies and arrived at a mean correlation for 14,553 parent–offspring pairs of 0.165 for SBP and 0.137 for DBP. Corresponding values for 11,839 sibling and DZ twin pairs were 0.235 (SBP) and 0.201 (DBP).

If, on the other hand, parents and their offspring are measured at the same age, a rise in parent–offspring correlations toward levels similar to sibling correlations is to be expected. This expectation was supported by data from Havlik et al. (1979), who measured SBP and DBP for 1141 parent pairs aged 48–51. At 20–30 years later, blood pressures for 2497 of their offspring were measured. At this time, the offspring were of ages similar to those of their parents when they were measured. Parent–offspring correlations ranged between 0.13 and 0.25 for SBP and between 0.17 and 0.22 for DBP. These ranges were quite similar to the sibling-pair correlations, which were between 0.17 and 0.23 (SBP) and between 0.19 and 0.24 (DBP).

An alternative explanation for the lower parent–offspring correlation compared to the sibling or DZ twin correlation could be the influence of genetic dominance (Tambs et al., 1992). The similarity between parent–offspring and sibling correlations in the study of Havlik et al. (1979) suggests, however, that dominance variation is not important.

Lower values for parent–offspring correlations also lead to lower h^2 estimates for blood pressure in family studies (which usually measure pairs of subjects at different ages) compared with twin studies. Heritability estimates from family studies range from 0.17 to 0.45 for SBP and from 0.15 to 0.52 for DBP (Iselius et al., 1983; Hunt et al., 1989; Rice, Vogler, Perusse, Bouchard, & Rao, 1989; Tambs et al., 1992), while estimates from twin studies range from 0.40 to 0.78 for SBP and from 0.32 to 0.76 for DBP (see Table 1).*

Province and Rao (1985) modeled genetic and environmental effects on SBP in nuclear families as a function of age. Before estimating environmental and genetic parameters, they used a standardization method to adjust for the effect of age on the mean and the variance; this standardization leaves temporal trends in familial resemblance intact. They found some evidence for a temporal trend in h^2, reaching maximum values of about 0.45 between 20 and 40 years of age. These results conflict with the results presented in Table 1, in which an age trend could not be detected. Thus, parent–offspring studies suggest that age may influence genetic effects on blood pressure, whereas no such effects are clear in twin data.

Two types of age-dependent effect could offer an explanation for the lower parent–offspring correlation compared to the sibling and DZ twin-pair correlations. First, the influence of nonshared environmental factors could increase with age. Such an increase, however, would lead to a lower h^2. Second, different genes could influence blood pressure in childhood than in adulthood. This possibility is still compatible with the results of Table 1, as h^2 can remain stable across time even though different genes are influential at different times. The latter possibility is supported by data from Tambs et al. (1993). In a Norwegian sample with 43,751 parent–offspring pairs, 19,140 pairs of siblings, and 169 pairs of twins, correlations between relatives decreased as age differences between these relatives increased. A model specifying age-specific genetic additive effects and unique environmental effects fitted the data well.

*Results of Theorell et al. (1979) are not considered because of deviant sample characteristics.

This model also estimated the extent to which genetic effects were age-specific. As an example, the expected correlations for SBP and DBP in relatives with an age difference of 40 years were calculated. For SBP, 62% of the genetic variance at, for example, age 20 and at age 60 is explained by genes that are common to both ages, and 38% is explained by age-specific genetic effects. The same values for DBP were 67% and 33%, respectively. The model used by Tambs et al. (1993) assumes invariant heritabilities for blood pressure throughout life. This assumption proved to be valid for SBP, whereas for DBP a very slight increase in h^2 was detected.

On the basis of their results from a study of twins and their parents, Sims, Hewitt, Kelly, Carroll, and Turner (1986) found that the assumption that the same genes act in young adulthood and middle age would require a *decrease* in heritability from 0.68 to 0.38 from young adulthood to middle age for DBP. This reduction, however, would need to be accompanied by an *increase* in the contribution of individual environmental factors that would account for an increase in phenotypic variance as people grow older. The same pattern of observations was seen for SBP (Sims, Carroll, Hewitt, & Turner, 1987). Samples in the studies of Sims et al. (1986, 1987) were relatively small (40 monozygotic [MZ], 45 dizygotic [DZ] male twin pairs, and their parents), and thus their results carry less weight than those from the very large Norwegian sample (Tambs et al., 1993). However, on the basis of the relatively small study of Sims et al. (1986, 1987), Hewitt, Carroll, Sims, and Eaves (1987) presented an alternative hypothesis for increases in variance of blood pressure across the life span that could be tested in longitudinal studies using genetically informative subjects. They proposed a developmental model in which genetic effects on blood pressure are largely the same (pleiotropic) at different points in time, but not cumulative throughout

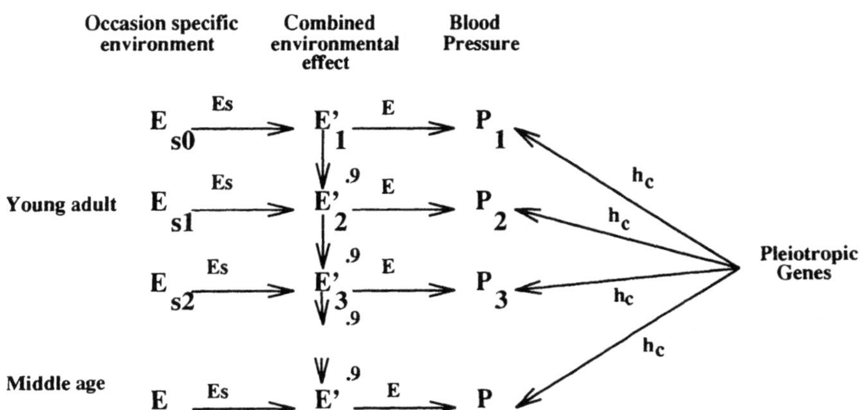

FIGURE 1. A developmental genetic model for adult blood pressure. Reprinted by permission from J. K. Hewitt, D. Carroll, J. Sims, and L. J. Eaves, "A Developmental Hypothesis for Adult Blood Pressure," *Acta Geneticae Medicae et Gemellologiae*, vol. 36, pp. 475–483. Copyright 1987 Assoociezione Instituto di Genetica Medica e Gemellologia Gregorio Mendel.

adulthood. Specific environmental influences are transient in their occurrence, but their impact is transmitted across occasions (Fig. 1).

Although it was restricted to twins aged in their late 40s and older and therefore did not cover the early adult to middle-aged period, the one reported longitudinal twin study (Colletto, Cardon, & Fulker, 1993)—which is discussed below—appears to disconfirm this hypothesis, finding little evidence for such a transmission or accumulation of environmental influences.

Blood Pressure Reactivity

The few parent–offspring studies reported to date will not permit a reliable conclusion about age-dependent genetic effects on blood pressure reactivity, and the comparison of parent–offspring with sibling or DZ twin correlations is difficult because of this paucity of studies. Moreover, the published studies to date have generally found low, or unstable, familial correlations, and quantitative genetic model-fitting has not been attempted. Matthews et al. (1988) conducted a study in which 145 families were measured using serial subtraction, mirror image tracing, and an isometric handgrip task. Only the correlation of SBP reactivity to the isometric handgrip was significant for parent–offspring as well as sibling pairs. Ditto (1987) investigated similarities in 36 sibling pairs (aged 18–21) in cardiovascular reactivity to four different stress tasks: a conceptual task, a mental arithmetic task, an isometric handgrip task, and a cold pressor task. Only one significant sibling correlation was found for DBP reactivity to the cold pressor task (0.40).

Ditto and France (1990) compared correlations in a group of 30 young (mean age: 26) and 30 middle-aged (mean age: 52) spouse pairs. Spouse correlations in blood pressure reactivity to a mental arithmetic and an isometric handgrip task were small and did not increase as a function of the number of years of living together. This suggests that the influence of a common environmental factor is small or even nonexistent, thereby offering support for the second conclusion of Turner and Hewitt (1992): that shared environment plays only a minor role in determining individual differences in blood pressure reactivity.

LONGITUDINAL STUDIES: BLOOD PRESSURE LEVELS

While we know of no longitudinal studies of genetic and environmental influences on blood pressure reactivity, two such studies have been reported for blood pressure levels. In such studies, variables are measured on repeated occasions, over an extended period, in the same subjects. A resulting matrix of correlations across occasions often conforms to what is referred to as the *simplex* pattern, in which correlations are maximal among adjoining occasions and decrease as the time between measurements increases. Such data may be described by autoregressive models in which some random change in the underlying phenotype, distinct from measurement errors, is introduced on each occasion. The underlying phenotype continually changes, making measurements from adjacent time periods more similar than those from more remote ones. This autoregressive simplex model can be generalized to the genetic analysis of longitudinal data, providing information on genetic stability across time (Boomsma & Molenaar, 1987).

Although they did not exploit these modern methods of time series analysis, Hanis, Sing, Clarke, and Schrott (1983) studied genetic and environmental influences on familial aggregation, coaggregation, and tracking (intraindividual correlation over time) of SBP and weight. The study started with a sample of 998 full sibs and first cousins from 261 families of whom 601 were measured on 4 occasions. Data from occasions 1 and 2 were analyzed together as Group 1, and those from times 3 and 4 were similarly analyzed as Group 2. Models were fit first to Group 1 and then to Group 2. With respect to tracking, 59% of the SBP-tracking correlation (0.33) and 60% of the weight-tracking correlation (0.89) were attributable to genetic effects. Relationships between SBP and weight remained stable over time. Two shortcomings of this study, however, should be noted: the short period of follow-up (4 years) and the subjects' young age. Subjects were measured during preadolescence (from 9.2 to 13.3 years), whereas it is likely that the development of blood pressure can be divided into a preadult and an adult phase (Hewitt et al., 1987).

During adulthood, the developmental genetics of blood pressure might be similar to that recently reported for the body mass index; although the overall heritability remains relatively constant from young adulthood on, there are nevertheless additional genetic influences acting in middle age independent of those that influence young adults (Fabsitz, Carmelli, & Hewitt, 1992). This possibility has been given support by the report of Colletto et al. (1993), who analyzed SBP and DBP for 254 MZ and 260 DZ male twin pairs assessed in middle age (mean age: 48 years) and again 9 years and 24 years later. Using a time series analysis of genetic and environmental components of variation, they found that shared family environmental effects were absent and that specific environmental influences were largely occasion-specific. In contrast, genetic influences were in part the same across adulthood (60% of genetic variation at the later ages was already detected in middle age) and in part age-specific (the remaining 40% of the genetic variation at later ages was unrelated to that expressed earlier). Despite these changing genetic influences, the estimated heritabilities remain relatively constant across ages at around 0.5.

A COMPROMISE: ESTIMATION OF AGE-DEPENDENT GENETIC AND ENVIRONMENTAL EFFECTS WITHOUT A LONGITUDINAL DESIGN

Though twin–family designs are advantageous in that they yield additional information from the parents of the twins (Sims et al., 1986, 1987; Eaves, Fulker, & Heath, 1989; Boomsma, 1992), a disadvantage of this design is the underlying assumption that the same genes are expressed in parents and their offspring. Stated differently, the assumption is that the correlation between genetic effects during childhood and adulthood equals unity. To test this assumption rigorously, longitudinal data from genetically informative subjects are needed. However, a less expensive and less time-consuming option is to extend the parent–offspring design to include, in addition to younger twins and their parents, a group of middle-aged twins of the same age as those parents (Stallings, Baker, & Boomsma, 1989).

This extended design, which includes young twins, their parents, and twins of the same age as the parents, can take account of the possibility that the correlation between genetic effects during adolescence and adulthood does not equal unity. The model can be written as:

$$R\text{po} = 0.5 \times R\text{g} \times H1 \times H2 \quad (1)$$

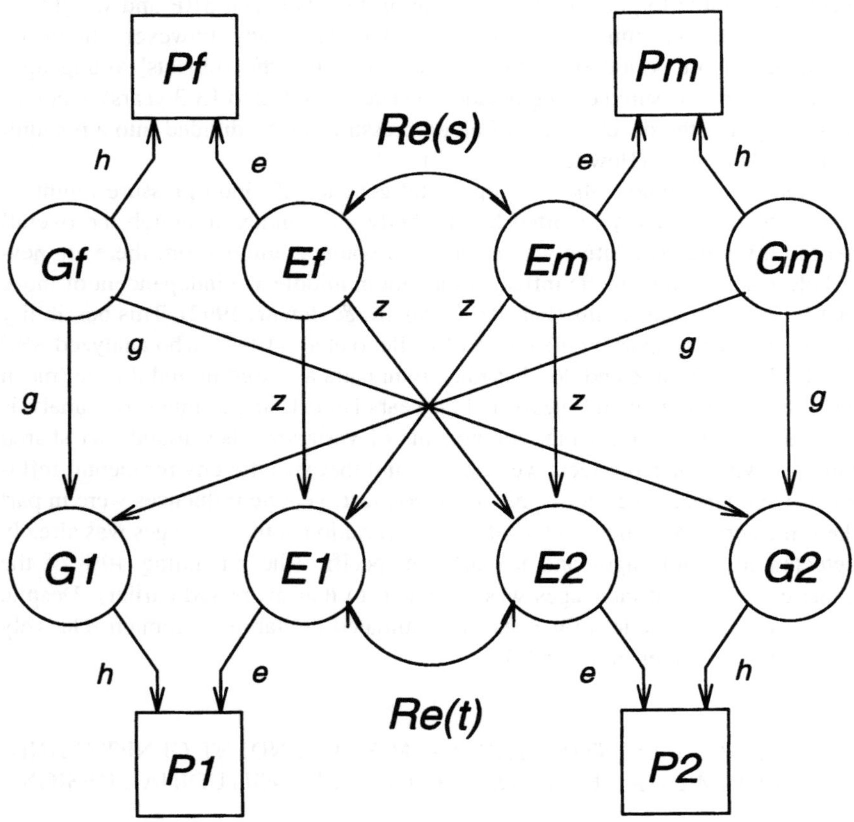

FIGURE 2. Parent–offspring model. (*P*) Observed phenotypes of father (*f*), mother (*m*), twin 1 (*1*) and twin 2 (*2*). (*G, E*) Latent genotype and total environment. (*g*) Modeling of genetic transmission from parents to offspring; (*z*) modeling of nongenetic transmission from total parental environment to offspring environment, in this case modeled to be equal for fathers and mothers. [*Re(s)*] Correlation between total environments of spouses; [*Re(t)*] correlation between residual environments of twins.

where *R*po is the parent–offspring correlation, *R*g is the genetic correlation across time (which is 1 if the same genes are expressed during adolescence and adulthood), *H*1 is the estimate of genetic influence during adolescence, and *H*2 is the estimate of genetic influence during middle age. *H*1 and *H*2 are equal to the square root of the corresponding heritabilities and can both be estimated with a univariate analysis for the adolescent and middle-aged twin groups, respectively. As *R*po is observed, *R*g is the only unknown left, which allows one to test whether *R*g differs from unity (Fig. 2). Figure 2 depicts the parent–offspring model as outlined by Boomsma, van den Bree, Orlebeke, and Molenaar (1989). In this model, the impact of total parental environment on offspring environment is considered, and the spouse correlation [*Re(s)*] is modeled as a correlation between total environments. The total environmental offspring correlation consists of a part that is accounted for by parental influences {through the environmental transmission parameter (*z*) and a part independent of the resemblance with their parents [*Re(t)*]}.

In the extended model, parameters are estimated by using information from 10 groups: 5 groups of young twins and their parents and 5 middle-aged twin groups, grouped according to the sex and zygosity of the twins. In this multigroup design, path coefficients for the parents equal those of the middle-aged twins. The genetic transmission from parents to offspring is modeled by the *g* coefficient. This *g* coefficient equals 0.5 *R*g and *g* equals 0.5 if the same genes are expressed in adolescence and adulthood {as *R*g equals 1 in that case [see equation (1)]}. A prerequisite for the application of this model is a significant parent–offspring resemblance, which implies significant heritabilities in both childhood and adulthood and a substantial genetic correlation across time [see equation (1)].

Blood Pressure Levels

This approach was adopted in the modeling of our own data. Data were combined from two different research projects. In the first project (Boomsma, 1992), SBP and DBP were measured in a group of adolescent twins and their parents. In the other project (which is still in progress), blood pressure data from twins in the same age range as the parents were collected. Some characteristics of the subject groups are shown in Table 3.

TABLE 3. Number of Pairs, Mean Age, Standard Deviations, and Age Range of Included Subject Groups

		Age	
Subject group	*N*	Mean (SD)	Range
Young twins	156	16.7 (2.0)	13–22
Parents	156	46.8 (6.3)	35–65
Middle-aged twins	200	44.4 (6.7)	35–62

TABLE 4. Number of Individuals and Mean Systolic and Diastolic Blood Pressures and Standard Deviations for Males and Females in the Three Subject Groups

	Males					Females				
Subject group	*N*	SBP	(SD)	DBP	(SD)	*N*	SBP	(SD)	DBP	(SD)
Young twins	160	119.8	(8.8)	65.8	(6.8)	152	114.9	(6.5)	67.6	(5.1)
Parents	156	128.2	(10.6)	80.7	(8.0)	156	125.7	(14.1)	77.6	(9.9)
Middle-aged twins	189	127.4	(10.8)	78.6	(8.7)	211	122.1	(13.5)	74.2	(10.3)

Blood pressure was measured three times by the Dinamap 845XT, using oscillometric techniques, while subjects rested and were comfortably seated in a sound-attenuated cabin for 8.5 minutes. In the first project only, this condition was repeated once. One average SBP value and one average DBP value for the rest condition were calculated in all groups. Mean values and standard deviations of SBP and DBP for males and females in the three subject groups are presented in Table 4. Both means and standard deviations of SBP and DBP are larger in the parents and the middle-aged twin group compared to the young twin group.

Before genetic modeling using LISREL VII (Jöreskog & Sörbom, 1988), blood pressure data were corrected for body weight. Maximum-likelihood estimates of parent–offspring correlations are shown in Table 5.

For SBP, the model in which the spouse correlation was set to zero fitted best. In the best-fitting model for DBP, the spouse correlation was set to zero and the four different parent–child correlations were set equal. Models in which parent–child correlations were set to zero fitted significantly less well for both SBP and DBP (Table 5), which means that there is a significant parent–offspring correlation.

As stated earlier, *H*1 and *H*2 (Equation [1]) can be estimated with a univariate analysis within the adolescent and middle-aged twin groups, respectively. Twin correlations of both groups are shown in Table 6. The pattern of correla-

TABLE 5. Models for Systolic and Diastolic Blood Pressure Rendering Maximum-Likelihood Estimates of Parent–Offspring Correlations[a]

Measure	*R*sp	*R*fs	*R*fd	*R*ms	*R*md	χ^2	*df*	*p*
SBP	0.06	0.17	−0.02	0.22	0.44	39.36	36	0.32
	0	**0.16**	**−0.04**	**0.21**	**0.44**	**39.88**	**37**	**0.34**
	0	0.19	0.19	0.19	0.19	55.41	40	0.05
	0	0	0	0	0	72.36	41	0.00
DBP	−0.04	0.08	0.13	0.30	0.26	25.54	36	0.90
	0	0.09	0.14	0.30	0.27	25.78	37	0.92
	0	**0.20**	**0.20**	**0.20**	**0.20**	**29.50**	**40**	**0.89**
	0	0	0	0	0	45.94	41	0.28

[a]Correlations: (*R*sp) spouse; (*R*fs) father–son; (*R*fd) father–daughter; (*R*ms) mother–son; (*R*md) mother–daughter. The best-fitting models are in **boldface** type.

TABLE 6. Twin Correlations for Systolic and Diastolic Blood Pressure in Both Young Twins and Middle-Aged Twins

	Young twins			Middle-aged twins		
Zygosity and sex[a]	Number of pairs	SBP	DBP	Number of pairs	SBP	DBP
MZF	33	0.51	0.49	47	0.50	0.59
DZF	29	0.33	0.40	39	0.42	0.28
MZM	35	0.49	0.61	43	0.44	0.44
DZM	31	0.53	0.56	32	0.16	0.23
DOS	28	0.37	0.32	39	0.23	0.21

[a]See the Table 1 footnote for abbreviations.

tions as shown in Table 6 is indicative of genetic influences on blood pressure, as MZ twin correlations are larger than DZ twin correlations. Standardized estimates from univariate analyses are shown in Table 7 for young and middle-aged twins.

For both young and middle-aged twins, models estimating additive genetic (G) and environmental (E) factors fitted beset. For both groups, a G and E model that allowed heritabilities to be different in males and females gave the best account of the data.

Under these conditions of a significant parent–offspring resemblance and significant heritabilities in both childhood and adulthood, it is possible to fit the extended model (Fig. 2) and estimate the value of the genetic transmission coefficient. Results are shown in Table 8.

TABLE 7. Standardized Estimates of Best-Fitting Univariate Models for Systolic and Diastolic Blood Pressure in Twins[a]

Measure	h^2_m	e^2_m	h^2_f	e^2_f	χ^2	*df*	*p*
			Young twins				
SBP[b]	0.49	0.51	0.66	0.34	7.80	12	0.80
DBP[c]	0.69	0.31	0.50	0.50	5.81	12	0.93
			Middle-aged twins				
SBP[d]	0.40	0.60	0.63	0.37	23.98	12	0.02
DBP[d]	0.42	0.58	0.61	0.39	7.84	12	0.80

[a]Abbreviations: (h^2_m) heritability for males; (e^2_m) unique environmental variance for males; (h^2_f) heritability for females; (e^2_f) unique environmental variance for females.
[b]For SBP, absolute genetic variance is equal for males and females.
[c]For DBP, absolute unique environmental variance is equal for males and females.
[d]For both SBP and DBP, absolute unique environmental variance is equal for males and females.

TABLE 8. Standardized Estimates of Best-Fitting Extended Models for Systolic and Diastolic Blood Pressure[a]

	Parents[b]				Offspring[c]							
Measure	h^2_m	e^2_m	h^2_f	e^2_f	h^2_m	e^2_m	h^2_f	e^2_f	g	χ^2	df	p
SBP												
$g = 0.5$	0.26	0.74	0.57	0.43	0.47	0.53	0.65	0.35	0.50	81.35	59	0.029
g free	0.25	0.75	0.62	0.38	0.49	0.51	0.66	0.34	0.38	79.66	58	0.031
DBP												
$g = 0.5$	0.29	0.71	0.55	0.45	0.69	0.31	0.50	0.50	0.50	42.31	59	0.950
g free	0.38	0.62	0.60	0.40	0.70	0.30	0.51	0.49	0.36	39.56	58	0.970

[a] See footnote *a* in Table 7 for abbreviations. The best-fitting model in which the genetic transmission coefficient (g) is fixed at 0.5 and the one in which g is estimated are shown.

[b] For both SBP and DBP, absolute unique environmental variance is equal for males and females.

[c] For SBP, absolute genetic variance is equal for males and females; for DBP, absolute unique environmental variance is equal for males and females.

Models that estimated the genetic transmission coefficient fitted slightly better than models in which g was fixed at 0.5 for both SBP and DBP. This difference, however, was not significant, which means that a model with a g equal to 0.5 cannot be rejected on the basis of these data. The possibility that the same genes are active in parents and their offspring can therefore not be excluded. Although not visible in Table 8 (which present only standardized estimates), the increase in SBP and DBP variance with age (see Table 4) was explained in the best-fitting models by an increase in both genetic and unique environmental variance for females and, for males, an increase in unique environmental variance only. As a consequence, h^2 in males is smaller in the combined parent/middle-aged twin group compared to the h^2 in the young twins.

These results accord with data from Sims et al. (1986, 1987), who studied a group of male twins and their parents and also suggested a decrease in h^2 caused by an increase in unique environmental variance from young adulthood to middle age. In the models in which g was estimated rather than fixed at 0.5, g was estimated to be 0.38 for SBP and 0.36 for DBP (but not significantly different from 0.5). This result means that the genetic correlation across time [Rg from equation (1)] equals 0.76 for SBP and 0.72 for DBP. The slightly lower values found by Tambs et al. (1993) (0.62 for SBP and 0.67 for DBP) might be explained by the larger age difference (40 years) in their example, compared to the age difference between parents and offspring in this study (30 years).

To conclude: The slightly better fit of models estimating a genetic transmission coefficient offered some evidence, but not definitive evidence, that different genes influence blood pressure in childhood and in adulthood. The decrease of h^2 in males with age could be explained by an increase in unique environmental variance. The h^2 in females did not change with age, as unique environmental and genetic variance increased proportionally.

TABLE 9. Number of Individuals and Mean Systolic and Diastolic Blood Pressure Reactivity to a Reaction Time and a Mental Arithmetic Task and Standard Deviations for Males and Females in the Three Subject Groups

Subject group	Males			Females		
	N	SBP (SD)	DBP (SD)	*N*	SBP (SD)	DBP (SD)
Young twins						
RT	160	6.5 (5.4)	4.7 (3.7)	152	5.4 (5.6)	3.8 (3.4)
MA	160	10.0 (6.9)	7.8 (4.7)	152	9.6 (7.9)	7.3 (4.4)
Parents						
RT	156	5.7 (6.5)	3.3 (3.7)	156	4.6 (6.7)	1.8 (3.8)
MA	156	12.5 (8.1)	6.7 (4.1)	156	11.1 (9.0)	5.0 (4.3)
Middle-aged twins						
RT	189	7.9 (8.5)	3.7 (5.9)	211	6.7 (9.0)	3.2 (5.4)
MA	189	11.1 (10.2)	6.5 (7.0)	211	9.7 (10.4)	6.5 (6.8)

Blood Pressure Reactivity

Subjects were exposed to two mental stress tasks: a choice reaction time (RT) task and a pressured mental arithmetic (MA) task (for a detailed description of these tasks, see Boomsma, van Baal, & Orlebeke, 1990). Each task lasted 8.5 minutes, during which blood pressure was measured 3 times. In the first project only, tasks were repeated once. For each task, one average SBP and one average DBP value were calculated in all groups. Blood pressure reactivity was calculated as the absolute difference between blood pressure level during the tasks and during the resting period. Mean values and standard deviations of SBP and DBP reactivity to both tasks are presented in Table 9.

In all subject groups and in both sexes, SBP and DBP reactivity during the

TABLE 10. Best-Fitting Models for Systolic and Diastolic Blood Pressure Reactivity to a Reaction Time and a Mental Arithmetic Task Rendering Maximum-Likelihood Estimates of Parent–Offspring Correlations[a]

Measure	*R*sp	*R*fs	*R*fd	*R*ms	*R*md	χ^2	*df*	*p*
SBP								
RT	0	0.11	0.11	0.11	0.11	29.05	40	0.90
MA	0	0.13	0.13	0.13	0.13	42.69	40	0.36
DBP								
RT	0	0	0	0	0	76.94	41	0.00
MA	0	0	0	0	0	53.77	41	0.09

[a]See the Table 5 footnote for correlation abbreviations.

TABLE 11. Twin Correlations for Systolic and Diastolic Blood Pressure Reactivity to a Reaction Time and a Mental Arithmetic Task in Young Twins and Middle-Aged Twins

	Young twins					Middle-aged twins				
		SBP		DBP			SBP		DBP	
Zygosity and sex[a]	Number of pairs	RT	MA	RT	MA	Number of pairs	RT	MA	RT	MA
MZF	33	0.37	0.33	0.07	0.27	47	0.15	0.33	0.36	0.16
DZF	29	0.41	0.49	0.25	0.41	39	−0.23	0.00	−0.15	0.09
MZM	35	0.30	0.51	−0.04	0.33	43	0.54	0.48	0.19	0.32
DZM	31	0.16	0.07	−0.07	0.06	32	0.03	−0.38	0.19	−0.03
DOS	28	0.35	0.07	0.24	0.12	39	0.29	0.29	0.04	−0.02

[a]See the Table 1 footnote for abbreviations.

MA task was greater than during the RT task. Blood pressure responses to the tasks in males were slightly higher than in females. DBP reactivity in both sexes and to both tasks was somewhat larger in the young twins compared to the parents and middle-aged twins. Such a pattern was absent for SBP reactivity.

Maximum-likelihood estimates of parent–offspring correlations are shown in Table 10. In the best-fitting model for SBP reactivity to both tasks, the spouse correlation was set to zero and the four different parent–child correlations were set equal. For DBP reactivity to both RT and MA tasks, models in which spouse and parent–child correlations were set to zero gave the most parsimonious account of the data. This result means that there is no significant parent–offspring correlation for DBP reactivity. Univariate analyses for blood pressure reactivity to both tasks were executed to estimate $H1$ and $H2$ [equation (1)] within the adolescent and middle-aged twins groups, respectively. Twin correlations of both groups are shown in Table 11. No clear overall pattern pointing to either genetic

TABLE 12. Standardized Estimates of Best-Fitting Univariate Models for Systolic and Diastolic Blood Pressure Reactivity to a Reaction Time and a Mental Arithmetic Task in Young Twins[a]

Measure	h^2	c^2	e^2	χ^2	*df*	*p*
SBP						
RT	—	0.31	0.69	3.94	13	0.99
MA	0.44	—	0.56	25.63	13	0.02
DBP						
RT	—	—	1.00	23.21	14	0.06
MA	0.38	—	0.62	12.98	13	0.45

[a]Abbreviations: (h^2) heritability; (c^2) shared environmental variance; (e^2) unique environmental variance.

TABLE 13. Best-Fitting Univariate Models for Systolic and Diastolic Blood Pressure Reactivity to a Reaction Time and a Mental Arithmetic Task in Middle-Aged Twins[a]

Measure	h^2_m	e^2_m	h^2_f	e^2_f	χ^2	*df*	*p*
SBP							
RT[b]	0.37	0.63	0.27	0.73	27.58	12	0.01
MA	0.36	0.64	0.36	0.64	26.63	13	0.01
DBP							
RT	0.23	0.77	0.23	0.77	23.12	13	0.04
MA	0.25	0.75	0.25	0.75	10.78	13	0.63

[a]See footnote *a* in Table 7 for abbreviations.
[b]For SBP reactivity to the RT task, absolute unique environmental variance is equal for males and females.

or common environmental influences could be detected in these correlations. Some correlations were even negative.

Results of univariate analyses are shown in Table 12 for young twins and in Table 13 for middle-aged twins. For the young twins (Table 12), best-fitting models for the two stress tasks (RT and MA) and the two reactivity measures (SBP and DBP reactivity) did not show a consistent pattern. For the SBP reactivity to the RT task, a model invoking shared environment (*C*) and nonshared environmental (*E*) factors fitted best. A model estimating *E* only gave the best fit for the DBP reactivity to the RT task. For SBP and DBP reactivity to the MA task, a *G* and *E* model fitted best to the data. Best-fitting models were thus different for the two tasks.

For middle-aged twins (Table 13), a *G* and *E* model gave the best fit in all four conditions. For SBP reactivity to the RT task, this model allowed heritabilities to be different in males and females.

As mentioned earlier, two conditions have to be met to be able to fit the extended model (Fig. 2) and estimate a genetic transmission coefficient. These requirements were not met: No significant parent–offspring correlations were found for DBP reactivity to RT and MA tasks, and no heritabilities were found for SBP and DBP reactivity to the RT task in the young twins. Only for SBP reactivity to the MA task was the parent–offspring correlation greater than zero and heritabilities found in childhood and in adulthood.

The picture is thus very far from clear for blood pressure reactivity. For SBP reactivity, a significant parent–offspring correlation was found; for DBP reactivity, no such correlation was found. Different models fitted best for different tasks, and for the same task, different models fitted best in young and in middle-aged twins.

DISCUSSION

This chapter has examined whether and how underlying genetic or environmental influences, or both, lead to stability or change across the life span in individual differences in blood pressure level and blood pressure reactivity to stress. Different types of genetic studies and models investigating changes with age of heritability of blood pressure level and reactivity were discussed to shed some light on this question.

In twin studies of blood pressure level, no age trend in h^2 could be detected. Findings in family studies of lower parent–offspring compared to sibling and DZ twin correlations indicate, however, that age may influence genetic or environmental effects on blood pressure level. This age dependency could take two forms: The influence of unique environmental factors could increase with age, or different genes could influence blood pressure in children and adults. For both possibilities, some evidence was found in the literature (Sims et al., 1986, 1987; Tambs et al., 1993). However, an increase of unique environmental variance in adulthood, without a commensurate increase in genetic variance, would lower the heritability estimate, and the lack of an age trend in h^2 in twin studies is inconsistent with this prediction. On the other hand, the twin data are not inconsistent with the hypothesis of genes switching on and off with age, because the overall influence of genes can remain stable even though different genes are responsible for the effect.

Modeling of our own data gave some evidence, but not definitive evidence, that different genes are active during childhood and adulthood. However, a fall in h^2 with age in males could be explained by a rise in unique environmental variance, thus supporting the other possible age effect. In females, because environmental and genetic variance increased proportionally, heritability did not change with age. It seems that at least in males, both age-dependent effects on SBP and DBP level could act simultaneously: Different genes influence blood pressure in children and adults, and unique environmental variance increases with age.

As for blood pressure reactivity, in twin studies, no age trend in h^2 could be detected either, and the few family studies reported did not allow any conclusions about age dependency.

Furthermore, the picture that arises from modeling of our own data is far from clear. A significant parent–offspring correlation was found for SBP reactivity, but not for DBP reactivity. Different models fitted best in young and middle-aged twins, and in young twins, the best-fitting models for both SBP and DBP reactivity were different for the two tasks. This finding is somewhat unexpected if it is assumed that responses to mental tasks represent the underlying general propensity of subjects to be reactive or not. The same task dependency of heritability estimates was observed by Ditto (1993): The SBP response to a concept formation task showed no heritability, whereas the response to a mental arithmetic task did.

A reason for these differences might be that the blood pressure response is a composite of several contributing mechanisms that may differ between tasks (see, for example, Turner [1994] and Chapter 5). In one task, vascular processes may be the main determinant of the blood pressure elevation, while cardiac involvement may be more prominent in another task. If the genetic contribution to cardiac and vascular reactivity is different, this difference will lead to different heritability estimates of the blood pressure response to these tasks. Studying the genetic contribution to blood pressure reactivity is, in fact, studying the combined genetic effects on several intermediate phenotypes such as vascular reactivity and baroreflex sensitivity. Future studies will have to measure the genetic contributions to the response of cardiac sympathetic (pre-ejection period) and parasympathetic indices [respiratory sinus arrythmia (Boomsma et al., 1990)] and to the response of indices of peripheral resistance and cardiac output. Such studies may reveal why heritability estimate for reactivity differ between stressors. Moreover, more knowledge of the genetics of these intermediate phenotypes will bring the genetic approach close to the crucial questions in hypertension research.

There is growing evidence that in contrast to earlier ideas, baroreceptor sensitivity and cardiac and vascular structural changes (all of which influence reactivity) have a genetic component (Weinstock, Weksler-Zangen, & Schorer-Apelbaum, 1986; Harrap, Van der Merwe, Griffin, MacPherson, & Lever, 1990; Unger & Rettig, 1990; Parmer, Cervenka, & Stone, 1992) and, moreover, are not merely a consequence of elevated pressure but may precede a rise in pressure and thus have an etiological role. The study of the genetics of blood pressure levels and of its age and sex dependency will benefit from measuring intermediate phenotypes in addition to blood pressure. Clearly, the genetic variance in blood pressure reflects the genetic variance of its determinants. A genetic contribution to stroke volume and peripheral resistance was shown by Bielen, Fagard, and Amery (1991). The age dependency of the genetic contribution to these parameters is of special interest because the contribution of the vasculature to high blood pressure is greater in older than in younger people (Lund-Johansen, 1977). By looking only at the age dependency of the genetic contribution to blood pressure, one may miss information about the genetics of mechanisms hypothesized to be involved in the etiology of high blood pressure. Our own data on these underlying mechanisms will become available in the near future.

A general problem in hypertension research, which applies to the study of age dependency of blood pressure, is the difficulty of determining whether abnormalities in blood-pressure-regulating mechanisms (in this case measured in different age groups) are causal factors for, or mere consequences of, elevated pressures. Moreover, conditions that have played a causal role at an early age may have disappeared when measured at a later age. For example an elevated cardiac output is a characteristic of some individuals with early borderline hypertension. With increasing age, the role of peripheral resistance as a determinant of blood pressure becomes more pronounced. Thus, in older people, one might be assessing the genetics of a consequence of elevated pressure instead of the genetics of a causal factor. To properly interpret the age dependency of the genetic contribution to blood pressure and its determinants, therefore, one must place the data in the framework of our knowledge of the etiology of hypertension development.

The study of the genetics of mechanisms involved in blood pressure regulation in young children might bring us closer to causal mechanisms. There is a considerable tracking of blood pressure levels from early to later childhood (Szklo, 1979), and blood pressure at young age is an important predictor of adult levels (Lauer & Clarke, 1989). Though moderately elevated blood pressure is a precursor of essential hypertension, only 20–30% of individuals will eventually reach the hypertensive range. A longitudinal twin study approach will allow examination of the difference in patterning of genetic and environmental factors between those who return to the normotensive range and those who develop hypertension.

The genetic autoregressive simplex model could be used in this respect to construct individual genetic and environmental profiles across time by means of a statistical technique known as Kalman filtering (Boomsma, Molenaar, & Dolan, 1991). Such individual profiles allow us to attribute individual phenotypic change to changes in the underlying genetic or environmental processes. Simulations have shown that these individual estimate can be reliably obtained. Estimation of genetic and environmental profiles across time would permit identification of sources of underlying deviant development in blood pressure for individual subjects.

Williams et al. (1990) have reviewed possible candidates for genetic and environmental causes of the development of hypertension. They expressed the hope that in the near future, biochemical tests and application of gene markers may provide methods for quantitatively assessing a person's risk for hypertension. A possible genetic marker for high blood pressure was identified in the parents of twins from our study: Individuals carrying the α_1-antitrypsin deficiency alleles S and Z had lower blood pressure during rest and stress. This finding replicate that from an Australian twin sample (Boomsma, Orlebeke, Martin, Frants, & Clark, 1991). Determining the balance between environmental and such predisposing genetic factors as early as possible is of utmost importance for preventive purposes, because high blood pressure, once established, is not easily reversed.

REFERENCES

Adams, M. A., Bobik, A., & Korner, P. I. (1989). Differential development of vascular and cardiac hypertrophy in genetic hypertension. Relation to sympathetic function. *Hypertension, 14,* 191–202.

Barnes, R. F., Raskind, M., Gumbrecht, G., & Halter, J. B. (1982). The effect of age on the plasma catecholamine response to mental stress in man. *Journal of Clinical Endocrinology and Metabolism, 54,* 64–69.

Bielen, E. C., Fagard, R. H., & Amery, A. K. (1991). Inheritance of blood pressure and haemodynamic phenotypes measured at rest and during supine dynamic exercise. *Journal of Hypertension, 9,* 655–663.

Boomsma, D. I. (1992). *Quantitative genetic analysis of cardiovascular risk factors in twins and their parents.* Thesis: Free University of Amsterdam. Enschede: Febodruk.

Boomsma, D. I., & Molenaar, P. C. M. (1987). The genetic analysis of repeated measures. I. Simplex models. *Behavior Genetics, 17,* 111–123.

Boomsma, D. I., Molenaar, P. C. M., & Dolan, C. V. (1991). Estimation of individual genetic and environmental profiles in longitudinal designs. *Behavior Genetics, 21,* 241–253.

Boomsma, D. I., Orlebeke, J. F., Martin, N. G., Frants, R. R., & Clark, P. (1991) Alpha-1-antitrypsin and blood pressure. *Lancet, 337,* 1547.

Boomsma, D. I., van Baal, G. C. M., & Orlebeke, J. F. (1990). Genetic influences on respiratory sinus arrhythmia across different task conditions. *Acta Geneticae Medicae et Gemellologiae, 39,* 181–191.

Boomsma, D. I., van den Bree, M. B., Orlebeke, J. F., & Molenaar, P. C. M. (1989). Resemblances of parents and twins in sport participation and heart rate. *Behavior Genetics, 19,* 123–141.

Carmelli, D., Ward, M. W., Reed, T., Grim, C. E., Harshfield, G. A., & Fabsitz, R. R. (1991). Genetic effects on cardiovascular responses to cold and mental activity in late adulthood. *American Journal of Hypertension, 4,* 239–244.

Colletto, G. M., Cardon, L. R., & Fulker, D. W. (1993). A genetic and environmental time series analysis of blood pressure in male twins. *Genetic Epidemiology, 10,* 533–538.

De Visser, D. C. (1994). Stress-reactivity in the Dutch Hypertension and Offspring Study: An epidemiological approach to the psychophysiology of early hypertension. Thesis: Erasmus University, Rotterdam.

Ditto, B. (1987). Sibling similarities in cardiovascular reactivity to stress. *Psychophysiology, 24,* 353–360.

Ditto, B. (1993). Familial influences on heart rate, blood pressure, and self-report anxiety responses to stress: Results from 100 twin pairs. *Psychophysiology, 30,* 635–645.

Ditto, B., & France, C. (1990). Similarities within young and middle-aged spouse pairs in behavioral and cardiovascular response to two experimental stressors. *Psychosomatic Medicine, 52,* 425–434.

Eaves, L. J., Fulker, D., & Heath, A. (1989). The effects of social homogamy and cultural inheritance on the covariances of twins and their parents: A LISREL model. *Behavior Genetics, 19,* 113–122.

Eaves, L. J., Last, K. A., Young, P. A., & Martin, N. G. (1978). Model-fitting approaches to the analysis of human behavior. *Heredity, 41,* 249–320.

Fabsitz, R. R., Carmelli, D., & Hewitt, J. K. (1992). Evidence for independent genetic influences on obesity in middle age. *International Journal of Obesity, 16,* 657–666.

Faucheux, B. A., Lille, F., Baulon, A., Landau, J., Dupuis, C., & Bourlière, F. (1989). Heart rate and blood pressure reactivity during active coping with a mental task in healthy 18- to 73-year-old subjects. *Gerontology, 35,* 19–30.

Feinleib, M., Garrison, R. J., Fabsitz, R., Christian, J. C., Hrubec, Z., Borhani, N. O., Kannel, W. B., Rosenman, R. H., Schwartz, J. T., & Wagner, J. O. (1977). The NHLBI twin study of cardiovascular disease risk factors: Methodology and summary of results. *American Journal of Epidemiology, 106,* 284–295.

Folkow, B., Hallback, M., Lundgren, Y., & Weiss, L. (1970). Background of increased flow resistance and vascular reactivity in spontaneously hypertensive rats. *Acta Physiologica Scandinavica, 80,* 93–106.

Folkow, B., & Svanborg, A. (1993). Physiology of cardiovascular aging. *Physiological Reviews, 73,* 725–763.

Furchtgott, E., & Busemeyer, J. K. (1979). Heart rate and skin conductances during cognitive processes as a function of age. *Journal of Gerontology, 34,* 183–190.

Garwood, M., Engel, B. T., & Capriotti, R. (1982). Autonomic nervous system functioning and aging: Response specificity. *Psychophysiology, 19,* 378–385.

Gintner, G. G., Hollandsworth, J. G., & Intrieri, R. C. (1986). Age differences in cardiovascular reactivity under active coping conditions. *Psychophysiology, 23,* 113–120.

Hanis, C. L., Sing, C. F., Clarke, W. R., & Schrott, H. G. (1983). Multivariate models for human genetic analysis: Aggregation, coaggregation and tracking of systolic blood pressure and weight. *American Journal of Human Genetics, 35,* 1196–1210.

Harrap, S. B., Van der Merwe, W. M., Griffin, S. A., MacPherson, F., & Lever, A. F. (1990). Brief angiotensin converting enzyme inhibitor treatment in young spontaneously hypertensive rats reduces blood pressure long-term. *Hypertension, 16,* 603–614.

Havlik, R. J., Garrison, R. J., Feinleib, M., Kannel, W. B., Castelli, W. P., & McNamara, P. M. (1979). Blood pressure aggregation in families. *American Journal of Epidemiology, 110,* 304–312.

Hewitt, J. K., Carroll, D., Sims, J., & Eaves, L. J. (1987). A developmental hypothesis for adult blood pressure. *Acta Geneticae Medicae et Gemellologiae, 36,* 475–483.

Hunt, S. C., Hasstedt, S. J., Kuida, H., Stults, B. M., Hopkins, P. N., & Williams, R. R. (1989). Genetic heritability and common environmental components of resting and stressed blood pressures, lipids and body mass index in Utah pedigrees and twins. *American Journal of Epidemiology, 129,* 625–638.

Iselius, L., Morton, N. E., & Rao, D. C. (1983). Family resemblance for blood pressure. *Human Heredity, 33,* 277–286.

Jöreskog, K. G., & Sörbom, D. (1988). *LISREL VII: A guide to the program and applications.* Chicago: SPSS Inc.

Kannel, W. B., (1976). Epidemiology of cerebrovascular disease. In R. Ross Russell (Ed.), *Cerebral arterial disease* (pp. 1–23). Edingburgh: Churchill Livingston.

Labarthe, D. R., Mueller, W. H., & Eissa, M. (1991). Blood pressure and obesity in childhood and adolescence: Epidemiological aspects. *Annals of Epidemiology, 1,* 337–345.

Lauer, R. M., Burns, T. L., Clarke, W. R., & Mahoney, L. T. (1991). Childhood predictors of future blood pressure. *Hypertension, 18(Supplement I),* I-74–I-84.

Lauer, R. M., & Clarke, W. R. (1989). Childhood risk factors for high adult blood pressure: The Muscatine study. *Pediatrics, 84,* 633–641.

Lerner, D. J., & Kannel, W. B. (1986). Patterns of coronary heart disease morbidity and mortality in the sexes: A 26-year follow-up of the Framingham population. *American Heart Journal, 111,* 383–390.

Levine, R. S., Hennekens, C. H., Perry, A., Cassady, J., Gelband, H., & Jesse, M. J. (1982). Genetic variance of blood pressure levels in infant twins. *American Journal of Epidemiology, 116,* 759–764.

Lund-Johansen, P. (1977). Central haemodynamics in essential hypertension. *Acta Medica Scandinavica Supplementum, 606,* 35–42.

Manuck, S. B., Kasprowicz, A. L., & Muldoon, M. F. (1990). Behaviorally-evoked cardiovascular reactivity and hypertension: Conceptual issues and potential associations. *Annals of Behavioral Medicine, 12,* 17–29.

Matthews, K. A., Manuck, S. B., Stoney, C. M., Rakaczky, C. J., McCann, B. S., Saab, P. G., Woodall, K. L., Block, D. R., Visintainer, P. F., & Engebretson, T. O. (1988). Familial aggregation of blood pressure and heart rate responses during behavioral stress. *Psychosomatic Medicine, 50,* 341–352.

Matthews, K. A., Meilahn, E., Kuller, L. H., Kelsey, S. F., Caggiula, A. W., & Wing, R. R. (1989). Menopause and risk factors for coronary heart disease. *New England Journal of Medicine, 321,* 641–646.

Matthews, K. A., & Stoney, C. M. (1988). Influences of sex and age on cardiovascular responses during stress. *Psychosomatic Medicine, 50,* 46–56.

Matthews, K. A., Woodall, K. L., & Allen, M. T. (1993). Cardiovascular reactivity to stress predicts future blood pressure status. *Hypertension, 22,* 479–485.

McIlhany, M. L., Shaffer, J. W., & Hines, E. A. (1975). The heritability of blood pressure: An investigation of 200 pairs of twins using the cold pressor test. *Johns Hopkins Medical Journal, 136,* 57–64.

Menkes, M. S., Matthews, K. A., Krantz, D. S., Lundberg, U., Mead, L. A., Quaqish, B., Lian,

K. Y., Thomas, C. B., & Pearson, T. A. (1989). Cardiovascular reactivity to the cold pressor test as a predictor of hypertension. *Hypertension, 14,* 524–530.

Molenaar, P. C. M., Boomsma, D. I., & Dolan, C. V. (1991). Genetic and environmental factors in a developmental perspective. In D. Magnusson, L. R. Bergman, G. Rudinger, & B. Thorestad (Eds.), *Problems and methods in longitudinal research: Stability and change* (pp. 250–273). Cambridge: Cambridge University Press.

Parmer, R. J., Cervenka, J. H., & Stone, R. A. (1992). Baroreflex sensitivity and heredity in essential hypertension. *Circulation, 85,* 497–503.

Plomin, R., DeFries, J. C., & McClearn, G. E. (1990). *Behavior genetics: A primer.* San Francisco: W. H. Freeman.

Pooling Project Research Group (1978). Relationship of blood pressure, serum cholesterol, smoking habit, relative weight and ECG abnormalities to incidence of major coronary events: Final report of the pooling project. *Journal of Chronic Diseases, 31,* 201–306.

Province, M. A., & Rao, D. C. (1985). A new model for the resolution of cultural and biological inheritance in the presence of temporal trends: Application to systolic blood pressure, *Genetic Epidemiology, 2,* 363–374.

Rice, T., Vogler, G. P., Perusse, L., Bouchard, C., & Rao, D. C. (1989). Cardiovascular risk factors in a French-Canadian population: Resolution of genetic and familial environmental effects on blood pressure using twins, adoptees, and extensive information on environmental correlates. *Genetic Epidemiology, 6,* 571–588.

Schieken, R. M., Eaves, L. J., Hewitt, J. K., Mosteller, M., Bodurtha, J. N., Moskowitz, W. B., & Nance, W. E. (1989). Univariate genetic analysis of blood pressure in children: The Medical College of Virginia twin study. *American Journal of Cardiology, 64,* 1333–1337.

Shapiro, A. P., Nicotero, J., & Schieb, E. T. (1968). Analysis of the variability of blood pressure, pulse rate, and catecholamine responsivity in identical and fraternal twins. *Psychosomatic Medicine, 30,* 506–520.

Sims, J., Carroll, D., Hewitt, J. K., & Turner, J. R. (1987). A family study of developmental effects upon blood pressure variation. *Acta Geneticae Medicae et Gemellologiae, 36,* 467–473.

Sims, J., Hewitt, J. K., Kelly, K. A., Carroll, D., & Turner, J. R. (1986). Familial and individual influences on blood pressure. *Acta Geneticae Medicae et Gemellologiae, 35,* 7–21.

Slattery, M. L., Bishop, D. T., French, T. K., Hunt, S. C., Meikle, A. W., & Williams, R. R. (1988). Lifestyle and blood pressure levels in male twins in Utah. *Genetic Epidemiology, 5,* 277–287.

Smith, T. W., Turner, C. W., Ford, M. H., Hunt, S. C., Barlow, G. K., Stults, B. M., & Williams, R. R. (1987). Blood pressure reactivity in adult male twins. *Health Psychology, 6,* 209–220.

Stallings, M. C., Baker, L. A., & Boomsma, D. I. (1989). Estimation of cultural transmission in extended twin designs. *Behavior Genetics, 19,* 777 (abstract).

Steptoe, A., & Ross, A. (1981). Psychophysiological reactivity and prediction of cardiovascular disorders. *Journal of Psychosomatic Research, 25,* 23–32.

Szklo, M. (1979). Epidemiologic patterns of blood pressure in children. *Epidemiological Review, 1,* 143–169.

Tambs, K., Eaves, L. J., Moum, T., Holmen, J., Neale, M. C., Naess, S., & Lund-Larsen, P. G. (1993). Age-specific genetic effects for blood pressure. *Hypertension, 22,* 789–795.

Tambs, K., Moum, T., Holmen, J., Eaves, L. J., Neale, M. C., Lund-Larsen, P. G., & Naess, S. (1992). Genetic and environmental effects on blood pressure in a Norwegian sample. *Genetic Epidemiology, 9,* 11–26.

Theorell, T., DeFaire, U., Schalling, D., Adamson, U., & Askevold, F. (1979). Personality traits and psychophysiological reactions to a stress interview in twins with varying degrees of coronary heart disease. *Journal of Psychosomatic Research, 23,* 89–99.

Turner, J. R. (1994). *Cardiovascular reactivity and stress: Patterns of physiological response.* New York: Plenum Press.

Turner, J. R., & Hewitt, J. K. (1992). Twin studies of cardiovascular response to psychological challenge: A review and suggested future directions. *Annals of Behavioral Medicine, 14,* 12–20.

Unger, T., & Rettig, R. (1990). Development of genetic hypertension: Is there a "critical phase"? *Hypertension, 16,* 615–616.

Valkenburg, H. A., Hofman, A., Klein, F., & Groustra, F. N. (1980). Een epidemiologisch onderzoek naar risico-indicatoren voor hart-en vaatziekten (EPOZ). I. Bloeddruk, serumcholesterolgehalte, quetelet-index en rookgewoonten in een open bevolking van 5 jaar en ouder. *Nederlands Tijdschrift voor Geneeskunde, 124,* 183–189.

van Lenthe, R. J., & Kemper, H. C. G. (1993). Risicofactoren voor hart- en vaatziekten en mogelijkheden voor preventie bij tieners: Het Amsterdams Groei en Gezondheid Onderzoek. *Hart Bulletin, 24,* 171–177.

Weinstock, M., Weksler-Zangen, S., & Schorer-Apelbaum, D. (1986). Genetic factors involved in the determination of baroreceptor heart rate sensitivity. *Journal of Hypertension, 4(Supplement 6),* s290–s292.

Williams, R. R., Hunt, S. C., Hasstedt, S. J., Hopkins, P. N., Wu, L. L., Berry, T. D., Stults, B. M., Barlow, G. K., & Kuida, H. (1990). Hypertension: Genetics and nutrition. In A. P. Simopoulos & B. Childs (Eds.), *World review of nutrition and dietetics,* Vol. 63, *Genetic variation and nutrition* (pp. 116–130). Basel: S. Karger.

PART FOUR

BODY WEIGHT, ADIPOSITY, AND EATING BEHAVIORS

CHAPTER SEVEN

Genetic Influences on Body Mass Index in Early Childhood

Lon R. Cardon

INTRODUCTION

It has become increasingly clear that hereditary factors play an important role in human fatness. Studies of adoptive families and twins reared together and apart have indicated a strong familial component to body fat that appears to be largely genetic in origin (Stunkard, Foch, & Hrubec, 1986; Stunkard, Harris, Pedersen, & McClearn, 1990; Price, Cadoret, & Stunkard, 1987; Price & Gottesman, 1991). Although such studies have typically focused on genetic contributions to fatness in adults, some evidence for a genetic etiology of body fat has also been obtained in studies of children (Brook, Huntley, & Slack, 1975; Price et al., 1990b).

Given that genetic influences have been indicated for body fat in both childhood and adulthood, a further domain of inquiry concerns the continuity of genetic effects over time. For weight-related characters, a positive correlation over successive measurements is expected, which may reflect continuous expression of genetic influences. Hewitt, Eaves, Neale, and Meyer (1988) noted that at least two mechanisms may account for such continuity in observations. The first mechanism, termed a *common factor* process, relates to situations in which common genetic or environmental factors determine measurements over multiple occasions, yielding consistently high or low observations. Deviations from the consistent levels may occur, but they are not expected to have any lasting effects

Lon R. Cardon • Sequana Therapeutics, 11099 North Torrey Pines Road, Suite 160, La Jolla, California 92037.

Behavior Genetic Approaches in Behavioral Medicine, edited by J. Rick Turner, Lon R. Cardon, and John K. Hewitt. Plenum Press, New York, 1995.

on future measurements. The second mechanism concerns the transmission of influences from occasion to occasion, in which the genetic effects at any one time (determined either by time-specific deviations or by previously expressed influences) have direct consequences for later observations. For body fat, this transmission model has important implications, because weight perturbations at a specific age may impact observed weight levels later in life.

To the extent that observed longitudinal continuity in body fat has a genetic basis, it would be useful to characterize, in childhood, individuals at risk for extreme fatness later in life. Rolland-Cachera et al. (1984, 1987) have described a childhood predictor of adult obesity, termed the *adiposity rebound,* that appears promising for identifying individuals susceptible to becoming obese and that requires only measurements of height and weight. A study by Cardon (1994) has examined the genetic basis of adiposity rebound and explored the extent to which those genetic factors that determine adiposity rebound in childhood relate to the genetic influences on body fat in adulthood.

In this chapter, we describe further results from analyses of the body mass index and adiposity rebound in participants of the Colorado Adoption Project (CAP), a study of genetic and environmental determinants of behavioral development. To our knowledge, studies of the CAP data represent the only reported investigations of the underlying processes that lead to continuity of fatness in early childhood. We also describe initial studies of the genetic commonality between childhood and adulthood. These studies have been conducted in the CAP by comparing relationships between the adopted and nonadopted children and their biological and adoptive parents, although we caution that analysis of parents and their children can provide only suggestive information concerning life-span development. Definitive information would necessitate a longitudinal study. Finally, we discuss the results concerning genetic and environmental etiologies of adiposity rebound and their ability to predict later fatness.

COLORADO ADOPTION PROJECT

The CAP is an ongoing longitudinal, prospective study in which 241 adoptive and 245 nonadoptive families have been evaluated at annual or semiannual intervals from birth to age 9. Adopted children in the project were separated from their biological parents a few days after birth and were placed in their adoptive homes within 1 month. Nonadoptive control families were matched to the adoptive families according to several criteria, including (1) age, education, and occupational status of the fathers; (2) gender of the adopted child; and (3) number of older children in the family. The sample is largely Caucasian (90% of the biological parents and 95% of the adoptive and control parents) and is largely representative of the Denver region for occupational status (Plomin & DeFries, 1985). Detailed descriptions of the CAP sample have been provided by Plomin, DeFries, and Fulker (1988).

The CAP study design has two main components: (1) parent–offspring relationships for inference of heritable effects transmitted vertically and (2) adopted and nonadopted siblings for investigations of shared genetic and environmental effects in a cohort of individuals of similar ages. The longitudinal aspect of the CAP permits analyses incorporating both the parent–offspring and sibling components of the project.

Siblings in this project are not ideal for some aspects of the analysis of differences in the control of developmental trends between genders because there are relatively few adoptive families in the CAP with same-sex sibling pairs, and such pairs are necessary to resolve sex-limited genetic effects (see Chapter 10). However, the parent–offspring analysis is suitable for exploration of gender effects because it involves only the parents and the proband children in the project, thus allowing simple tests of gender-based differences in parent–offspring relationships. The CAP children are represented roughly equally in numbers of males (53%) and females (47%).

The CAP protocol includes assessments of height and weight at every measurement occasion (annually from birth through age 7 and at age 9), which may be used to create an index of body fat, the body mass index (BMI). The BMI, calculated as weight (in kilograms) divided by the square of height (in meters), is expected to be a measure of weight that is largely independent of height. Consequently, the BMI is often described as a measure of fatness, adiposity, or obesity. Although the BMI, being a ratio of component measurements, has some inherent technical problems (Tanner, 1949), the scale is a useful approximation of laboratory measurements of body fat (Price, Ness, & Laskarzewski, 1990a). Analyses of the CAP data using multivariate methods applied to height and weight data have revealed results similar to those obtained using the BMI (Fulker, 1991).

MODELS OF DEVELOPMENT

The common factor and transmission models were originally formulated for biometrical twin analysis by Eaves, Long, and Heath (1986) and were subsequently described for adoption designs by Phillips and Fulker (1989). Some statistical properties of the models have been described by Hewitt et al. (1988). A path diagram encompassing both processes is shown in Fig. 1. In this diagram, longitudinal BMI values for an individual [shown as phenotypes (P)] are determined by an underlying common factor (F_c) with factor loadings c_n, persistent effects from previous occasions, g_n, and occasion-specific effects, s_n. In the application of this model, each of the parameters is specified for genetic influences, shared environmental influences, and nonshared environmental influences (Neale & Cardon, 1992). Thus, at any particular age ($n>0$), observed BMI ratios may result from constant genetic and environmental effects, persistent effects from the previous age, or age-specific influences. Evidence for a common factor

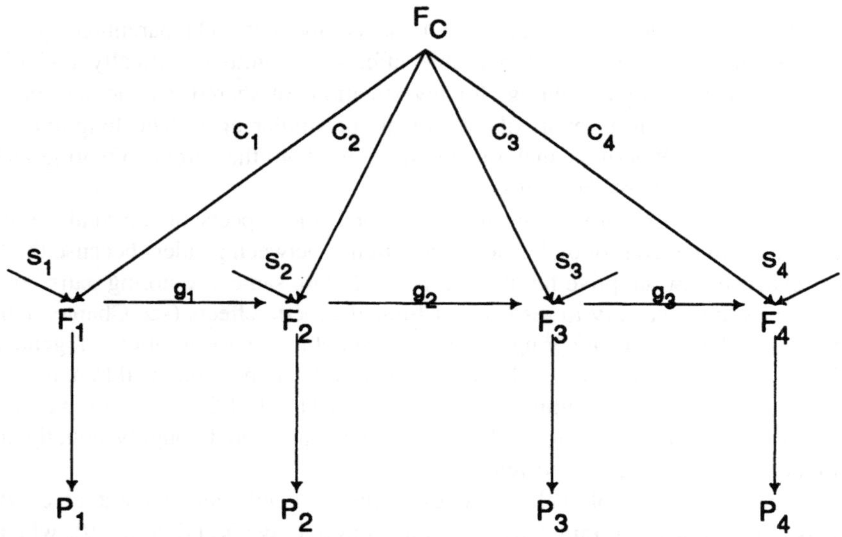

FIGURE 1. Developmental path model of genetic and environmental sources of covariance for one individual (see the text for explanation).

mechanism determining variation in BMI values is shown when all transmission parameters may be fixed at 0.0, whereas the transmission model is supported when all common factor loadings are set to 0.0.

PARENT–OFFSPRING MODELS

The path model employed for parent–offspring analyses in the CAP is presented in Fig. 2 for adoptive and biological families (Fulker, DeFries, & Plomin, 1988). In this model, BMI values are represented by the phenotypes P_{BM}, P_{BF}, P_{AM}, P_{AF}, and P_{AO} for biological mothers and fathers, adoptive mothers and fathers, and the adopted offspring, respectively. Hereditary influences are transmitted from the biological parents to the offspring ($1/2$) and environmental effects from the adoptive parents to the offspring (paths m and f). Assortative mating, or correlations between spouses for body fat, is represented by conditional paths (Carey, 1986) p and q for biological and adoptive parents. Selective placement is modeled using conditional paths x_{1-4}. A useful feature of the parent–offspring design is that it permits estimation of the correlation between genetic and environmental influences. This correlation is denoted as s in Fig. 2. The heritability (h^2) and environmentality (e^2) are equated for adults and children, thus providing estimates of shared genetic and environmental resemblance between parents and children.

Plomin (1986) has noted that since *h* is equated for parents and children in this model, our estimate of h^2 actually represents the product $h_C R_G h_A$, where R_G is the correlation between genetic influences in childhood (h_C) and adulthood (h_A). Since studies of adult twins and adoptees have yielded h^2_A estimates in the range of 0.40–0.60 [see Chapter 8 and Price et al. (1987)], estimates of h^2 in this model that are similar to these values would indicate a high degree of genetic resemblance between childhood and adulthood or greater childhood heritability, whereas lower estimates would indicate reduced genetic resemblance or reduced childhood h^2.

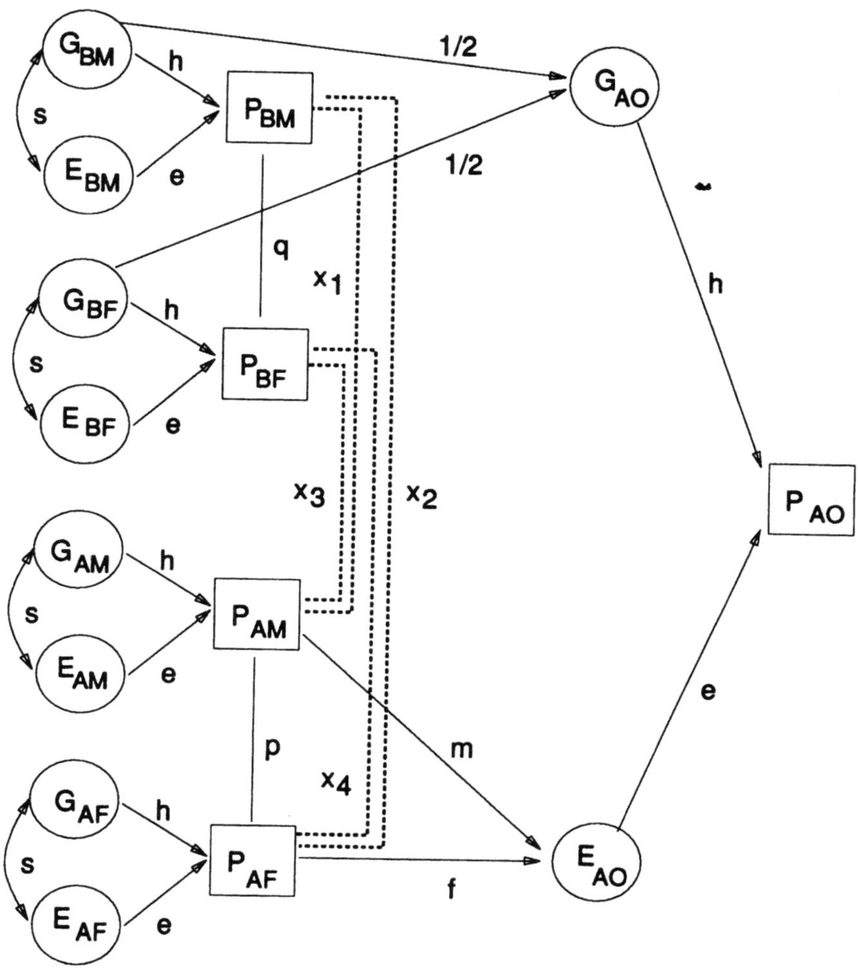

FIGURE 2. Model of parent–offspring resemblance. Reprinted by permission from D. W. Fulker, J. C. DeFries, and R. Plomin, "Genetic Influence on General Mental Ability Increases between Infancy and Middle Childhood," *Nature*, vol. 336, pp. 367–369. Copyright 1988 Macmillan Magazines Ltd.

CONTINUITY OF BODY MASS INDEX IN THE COLORADO ADOPTION PROJECT

Estimates of the heritability of BMI from parent–offspring and sibling analyses in the CAP are presented in Table 1. In the CAP siblings, BMI appears quite heritable throughout early childhood, with estimates of heritability ranging from 0.32 to 1.00 (the highest estimate of 1.00 is presumably due to a small sample size at age 9) (see Cardon, 1994). Moreover, continuity of body fat seems to result entirely from genetic factors, which follow the transmission pattern of direct genetic impact on subsequent occasions from previous measurements. This pattern of influence is shown in Fig. 3. Age-specific environmental effects exert a substantial effect on body fat at nearly all ages, but do not have a lasting influence over time.

In Fig. 3, several estimates of the transmission parameters are close to 1.0, indicating that the genetic influences at several of the ages persist in entirety to subsequent ages. These genetic effects are augmented by fresh genetic variation at virtually every age examined, suggesting that the genetic component of body fat has elements of both continuity and change during childhood development. However, Fig. 3 also shows that some of the transmission parameters are less than 1.0, indicating genetic effects that dissipate over time, or, when close to 0.0, indicating lack of genetic continuity. For example, genetic influences on BMI at birth, although substantial in magnitude, do not show the pattern of persistence apparent at other ages. Most of the genetic continuity in BMI does not appear until age 1, after which a high degree of continuity is observed through age 9.

Analyses of parents and their offspring indicate heritabilities similar to those from the analyses of the siblings after about age 2, but very different from those obtained in infancy. The heritabilities in Table 1 show estimates ranging from 0.01 to 0.09 from birth to age 2, but considerably increased heritabilities at ages 3 and above. At about age 4, the heritabilities appear to stabilize and resemble those noted in analysis of the sibling data. There is no suggestion of genotype–

TABLE 1. Heritability Estimates for Body Mass Index from Birth to Age 9

	Age (years)								
Study group	0	1	2	3	4	5	6	7	9
Siblings									
h^2	0.88	0.43	0.53	0.64	0.52	0.32	0.39	0.58	1.00
e^2	0.12	0.57	0.47	0.36	0.48	0.68	0.61	0.42	0.00
Parent–offspring									
h^2	0.09	0.01	0.09	0.37	0.52	0.38	0.55	0.38	0.57
e^2	0.91	0.99	0.91	0.69	0.60	0.68	0.58	0.65	0.51

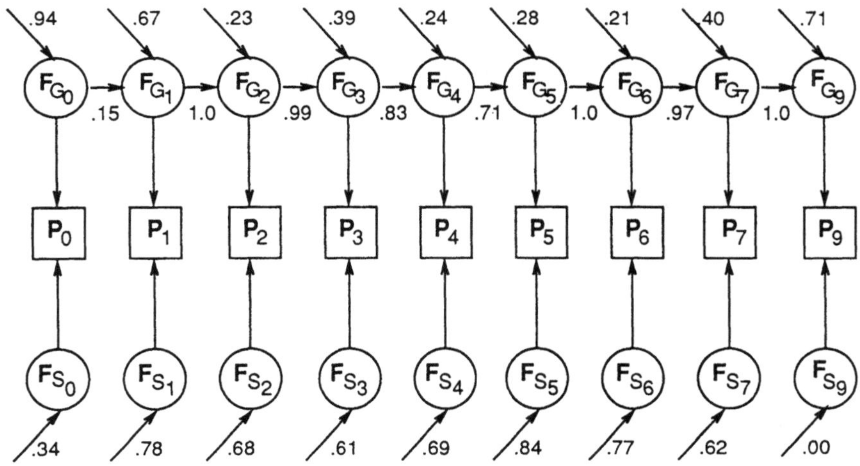

FIGURE 3. Final developmental model of body fat in siblings. BMI measurements are shown as phenotypes (P), with genetic (F_G) and within-family (F_S) influences.

environment correlation in these data. In conjunction with the sibling results, these findings indicate that although the genetic determinants of body fat in infancy relate to those in later childhood, they do not persist all the way into adulthood. The genetic effects that arise later in childhood, however, do continue to determine body fat in adult measurements.

We noted previously that the model employed for the parent–offspring analyses yields h^2 estimates that actually reflect the product of childhood and adult genetic effects (h_C and h_A) and the genetic correlation between them (R_G). Assuming an intermediate estimate of 0.60 for adult heritability of BMI from the twin and family studies reviewed in Chapter 8, the genetic correlation between adult and childhood BMI at each age may be estimated from the present sibling (${h^2}_C$) and parent–offspring (${h^2}_{p\text{-}o}$) estimates:

$$R_G = h^2_{p-o} / \sqrt{0.60 h^2_C}$$

Calculation of this quantity yields R_G estimates of 0.11, 0.02, 0.14, 0.52, 0.81, 0.75, 0.99, 0.56, and 0.64 between the various childhood ages (birth, 1–7, 9) and adulthood. These correlations again show that genetic effects on BMI assessments at age 3 or later relate much more strongly to adult body fat than do earlier effects. The correlations also indicate a striking degree of persistence of genetic effects from childhood to adulthood. For example, the correlation of 0.56 between age 7 and adulthood is equivalent to a transmission parameter of 0.976 between each intervening year (assuming an adult age of 30 years, a childhood h^2 of 0.58, and an adult h^2 of 0.60).

ADIPOSITY REBOUND

Adiposity rebound refers to the onset of a rapid growth in body fat that typically occurs at about 6 years of age. The growth trend may be seen in Fig. 4, in which the BMI means illustrate a decreasing, then increasing, trend between ages 2 and 9. For an individual, the inflection point at which BMI begins to rise is his or her adiposity rebound point. Rolland-Cachera et al. (1984) have described three general rebound categories for use in prediction of adult fatness: advanced (before age 5½), average (age 5½–7), and delayed (after age 7). They noted that children who exhibit the rebound early tend to have much greater body fat as adults, whereas children with middle and late rebound tend toward average

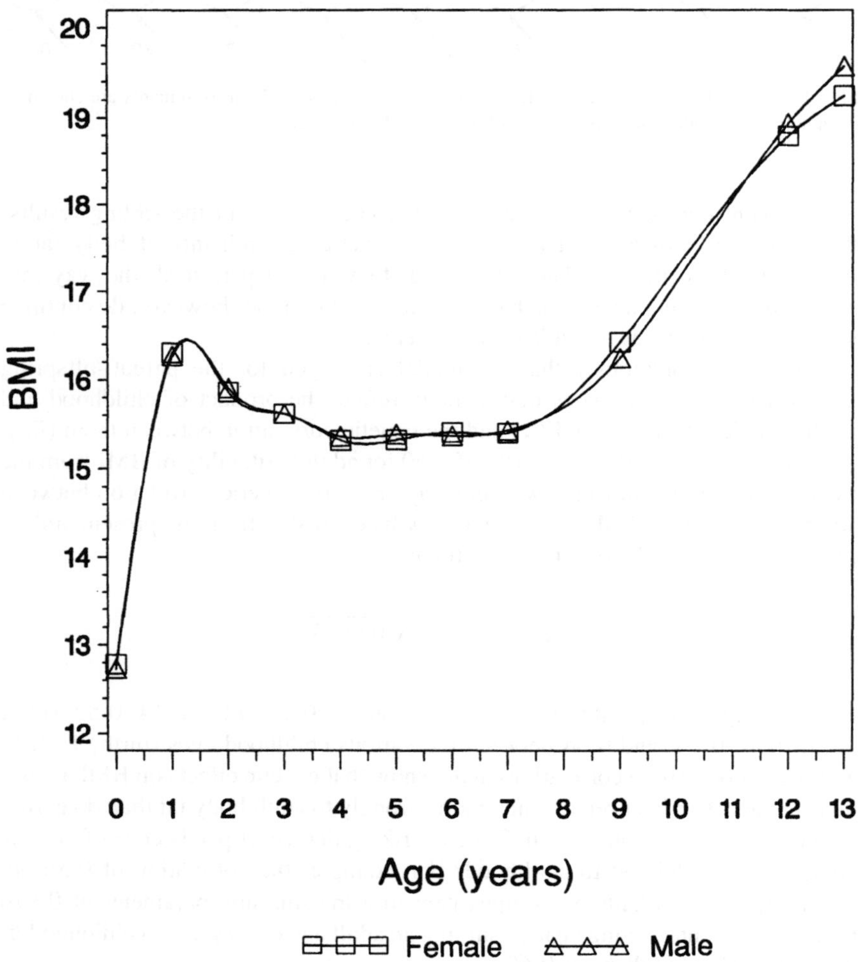

FIGURE 4. BMI means for adopted and nonadopted children in the Colorado Adoption Project.

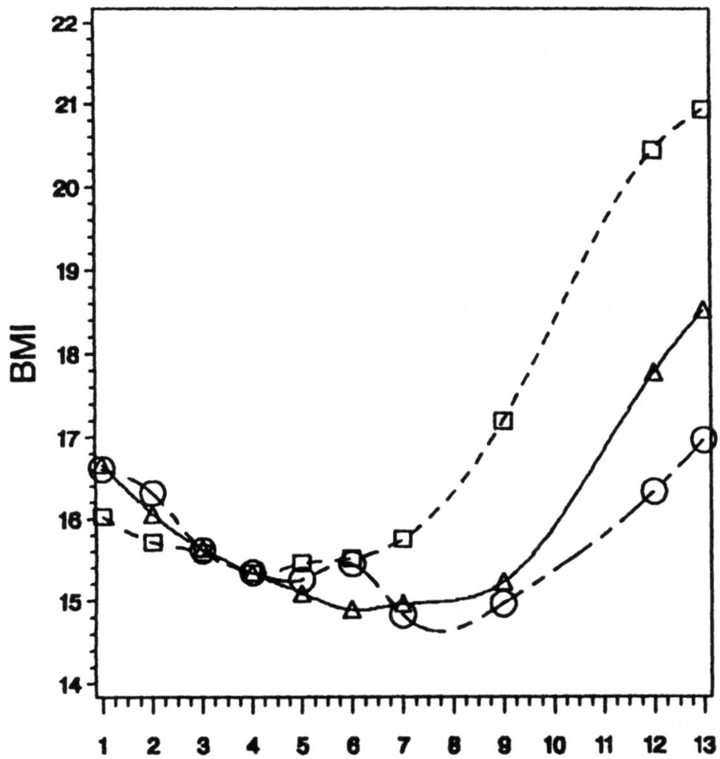

FIGURE 5. Adiposity rebound in the Colorado Adoption Project. BMI means are plotted for three categories of adiposity rebound: (□) early; (△) middle; (○) late.

fat levels. The adiposity rebound has been calculated for every child in the CAP sample. The children have also been classified as having advanced, average, or delayed rebound using the criteria of Rolland-Cachera et al. (1984).

The mean BMI values for children with advanced, average, and delayed adiposity rebound are shown in Fig. 5. It can be seen that the CAP data exhibit the pattern of effects expected from the earlier studies of adiposity rebound: Children with earlier rebound ages have higher BMI values later in development, and those with delayed rebound ages manifest lower indices.

Applications of standard regression procedures to estimate heritability (DeFries & Fulker, 1985, 1988) yield an estimate of shared environmentality (c^2) close to zero (-0.03) and an h^2 estimate of 0.37. This h^2 estimate is similar to that shown for the BMI itself, from both sibling and parent–offspring analyses. Additional analyses (Cardon, 1994) have shown that advanced, average, and delayed rebound are not differentially influenced by genetic factors.

Application of the parent–offspring model in Fig. 2 to parental BMI measures and offspring adiposity rebounds provides a means to investigate genetic effects common to adult BMI and childhood rebound. In the CAP data, this model yields an h^2 estimate of 0.27, suggesting that the genetic influences on adiposity rebound are related to those genetic factors that determine adult body fat, but that the relationship is not quite as strong as that of the BMI.

This parent–offspring h^2 estimate of 0.27 for adiposity rebound suggests that the predictability of adiposity rebound may result from similar genetic effects on adiposity rebound and adult fatness. Using the methods outlined above, the genetic correlation between adiposity rebound and adult BMI is 0.57, indicating a strong genetic relationship between the two measurements. This finding of shared genetic resemblance, combined with the results of previous studies showing that adiposity rebound may be used to identify individuals at risk for extreme fatness later in life (Rolland-Cachera et al., 1984, 1987), strongly suggests that this measure may be an important epidemiological indicator of adult obesity.

CONCLUSIONS

Results from the Colorado Adoption Project replicate previous studies in suggesting a substantial genetic component to fatness; they additionally show that continuity in body fat throughout childhood is determined largely by hereditary factors. The results support a dynamic developmental process of genetic persistence in which new genetic variation is accumulated each year, rather than a static process governed by a common genetic disposition. Only the genetic effects at birth appear discontinuous with those at later years. These findings indicate that novel genetic effects at early childhood ages, triggered perhaps in response to environmental stimuli, have measurable effects on body fat later in childhood. This conclusion is supported by the independent analyses of siblings and of parents and their offspring, suggesting that intervention directed at altering body fat in childhood may have benefits at later ages.

Childhood body fat is also determined by environmental influences. However, these effects, which comprise transient weight fluctuations arising from such things as illness and sudden periodic dieting, do not appear to have lasting consequences for body fat. Consistent observations of high or low body fat in childhood appear to result primarily from genetic factors.

Additional analyses of the timing of the adiposity rebound, focusing more specifically on obesity rather than on general body fat, supported findings from epidemiological studies and indicated heritable variation for this predictive trait. Chapter 8 takes up the question of the ultimate extent and nature of the genetic influences on BMI and obesity as we move from childhood, the focus of this chapter, into adult life.

REFERENCES

Brook, C., Huntley, R., & Slack, J. (1975). Influence of heredity and environment in determination of skinfold thickness in children. *British Medical Journal, 2*, 719–721.

Cardon, L. R. (1994). Height, weight, and obesity. In J. C. DeFries, R. Plomin, & D. W. Fulker (Eds), *Nature and nurture during middle childhood*. London: Blackwell.

Carey, G. (1986). A general multivariate approach to linear modeling in human genetics. *American Journal of Human Genetics, 39*, 775–786.

DeFries, J. C., & Fulker, D. W. (1985). Multiple regression analysis of twin data: Etiology of deviant scores versus individual differences. *Behavior Genetics, 15*, 467–473.

DeFries, J. C., & Fulker, D. W. (1988). Multiple regression analysis of twin data: Etiology of deviant scores versus individual differences. *Acta Geneticae Medicae et Gemellologiae, 37*, 205–216.

Eaves, L. J., Long, J., & Heath, A. C. (1986). A theory of developmental change in quantitative phenotypes applied to cognitive development. *Behavior Genetics, 16*, 143–162.

Fulker, D. W. (1991). Paper presented at the 17th conference of the Society for Behavioral Medicine, Chicago.

Fulker, D. W., DeFries, J. C., & Plomin, R. (1988). Genetic influence on general mental ability increases between infancy and middle childhood. *Nature, 336*, 767–769.

Hewitt, J. K., Eaves, L. J., Neale, M. C., & Meyer, J. M. (1988). Resolving causes of developmental continuity or "tracking." I. Longitudinal twin studies during growth. *Behavior Genetics, 18*, 122–151.

Neale, M. C., & Cardon, L. R. (1992). Methodology for genetic studies of twins and families. Dordrecht, The Netherlands: Kluwer Academic Publishers.

Phillips, K., & Fulker, D. W. (1989). Quantitative genetic analysis of longitudinal trends in adoption designs with application to IQ in the Colorado Adoption Project. *Behavior Genetics, 19*, 621–658.

Plomin, R. (1986). Multivariate analysis and developmental behavioral genetics: Developmental change as well as continuity. *Behavior Genetics, 16*, 25–43.

Plomin, R., & DeFries, J. C. (1985). *Origins of individual differences in infancy: The Colorado Adoption Project*. Orlando, FL: Academic Press.

Plomin, R., DeFries, J. C., & Fulker, D. W. (1988). *Nature and nurture during infancy and early childhood*. New York: Cambridge University Press.

Price, R. A., Cadoret, R. J., & Stunkard, A. J. (1987). Genetic contributions to human fatness: An adoption study. *American Journal of Psychiatry, 144*, 1003–1008.

Price, R. A., & Gottesman, I. I. (1991). Body fat in identical twins reared apart: Roles for genes and environment. *Behavior Genetics, 21*, 1–7.

Price, R. A., Ness, R., & Laskarzewski, P. (1990a). Common major gene inheritance of extreme overweight. *Human Biology, 14*, 747–765.

Price, R. A., Stunkard, A. J., Ness, R., Wadden, T., Heshka, S., Kanders, B., & Cormillot, A. (1990b). Childhood onset obesity has high familial risk. *International Journal of Obesity, 14*, 185–195.

Rolland-Cachera, M. F., Deheeger, M., Bellisle, F., Semp'e, M., Guilloud-Bataille, M., & Patois, E. (1984). Adiposity rebound in children: A simple indicator for predicting obesity. *American Journal of Clinical Nutrition, 39*, 129–135.

Rolland-Cachera, M. F., Deheeger, M., Guilloud-Bataille, M., Avons, P., Patois, E., & Semp'e, M. (1987). Tracking the development of adiposity from one month of age to adulthood. *Annals of Human Biology, 14*, 219–229.

Stunkard, A. J., Foch, T. T., & Hrubec, Z. (1986). A twin study of human obesity. *Journal of the American Medical Association, 256*, 51–54.

Stunkard, A. J., Harris, J. R., Pedersen, N. L., & McClearn, G. E. (1990). The body-mass index of twins who have been reared apart. *New England Journal of Medicine, 322*, 1483–1487.

Tanner, J. M. (1949). Fallacy of per-weight and per-surface area standards, and their relation to spurious correlation. *Journal of Applied Physiology, 2*, 1.

CHAPTER EIGHT

Genetic Studies of Obesity across the Life Span

JOANNE M. MEYER

INTRODUCTION

Obesity and correlated aspects of body fat distribution and patterning are established risk factors for several adult clinical disorders, including non-insulin-dependent diabetes, hypertension, atherosclerosis, and some cancers (Van Ittallie & Abraham, 1985; Barrett-Connor, 1985; Bray, 1985; Baumgartner, Roche, Chumlea, Siervogel, & Glueck, 1987; Blair, Habicht, Sims, Sylvester, & Abraham, 1984; Donahue, Abbot, Bloom, Reed, & Katsuhiko-Yano, 1987). Understanding the etiology of obesity is a necessary prerequisite for strategies aimed at reducing, and consequently ameliorating the effects of, excess weight. It has proven difficult, however, to obtain a thorough understanding of this etiology. The etiology is undoubtedly complex, and there is unlikely a single metabolic or behavioral explanation of individual differences in adiposity. Instead, the current evidence suggests that obesity is a consequence of *both* inherited (genetic) *and* acquired (environmental) aspects of caloric intake and energy expenditure. These factors most likely differ in their importance over an individual's lifetime.

Data from relatives, collected under twin, adoption, or parent–offspring designs, have made their greatest contribution to understanding the etiology of obesity by broadly characterizing the importance of genetic and environmental effects on the phenotype. This characterization has resulted in cross-sectional descriptions—How do genes and environment explain variation in obesity at any

JOANNE M. MEYER • Department of Human Genetics, Medical College of Virginia, Virginia Commonwealth University, Richmond, Virginia 23298.

Behavior Genetic Approaches in Behavioral Medicine, edited by J. Rick Turner, Lon R. Cardon, and John K. Hewitt. Plenum Press, New York, 1995.

one time?—and long-term prediction—What are the parental correlates of childhood obesity? Does obesity track over time? Why?. Following on Chapter Seven, which examined adoption study data from early childhood, the primary focus of this chapter is to summarize the available genetically informative data on a class of variables related to adiposity in later childhood and adulthood. We then discuss strategies for future analyses, suggested by the current literature, that aim to identify specific genetic risk factors for obesity.

GENETIC DESIGNS: QUESTIONS THEY CAN ANSWER ABOUT INDIVIDUAL DIFFERENCES IN OBESITY

A variety of constellations of biological and adoptive relative pairs, studied both cross-sectionally and longitudinally, have been used to investigate the familial nature of obesity. Each of the genetic designs has characteristic benefits that should be kept in mind when reviewing the published data or selecting a strategy for data collection. To summarize these benefits, we list in Table 1 questions that the different designs can answer about individual differences in obesity. For the purposes of presentation and discussion, we have made a distinction between cross-sectional and longitudinal studies.

A cross-sectional twin design refers to the comparison of monozygotic

TABLE 1. Study Designs and Questions They Can Answer about Familial Influences on Obesity

	Cross-sectional			Longitudinal		
Question	Twin	Parent–offspring	Adoption	Individual	Twin	Adoption
Is child obesity related to parent obesity?	—	X	X	—	—	X
Is familial similarity due to genes or to environment?	X	—	X	—	—	X
Does obesity heritability change over time?	X	—	—	—	X	X
Does individual obesity track over time?	—	—	—	X	X	X
How do genes explain tracking?	—	—	—	—	X	X
Do environmental influences persist over time?	—	—	—	—	X	X
Are new genes expressed over time?	—	X[a]	—	—	X	X

[a]When supplemented with twin data (see the text for a discussion).

(MZ) and dizygotic (DZ) twins, who may have been reared together or apart. The comparison of twins reared apart is preferable, since a rearing environment for MZ twins that overemphasizes exactly identical treatment cannot then influence twin similarity and inflate estimates of heritability (Price & Gottesman, 1991; Stunkard, Harris, Pedersen, & McClearn, 1990). However, it is difficult to obtain data on numbers of reared-apart twins large enough to have sufficient statistical power to detect genetic influences of moderate size. In contrast, the comparison of reared-together twins is more easily accomplished and can answer questions about familial similarity of obesity—Is it due to shared genes, shared environment or both?—in the absence of a special MZ twinning effect (see Chapter Two). Furthermore, when sufficient cross-sectional data are available to allow grouping of twins into restricted age ranges, questions regarding developmental changes in the heritability of obesity can be addressed (cf. Korkeila, Kaprio, Rissanen, & Koskenvuo, 1991).

A parent–offspring study refers to data on biological mother–child or father–child pairs who have resided in the same household while the child was being reared. Parent–offspring data on adult–child pairings are more common than those on adult–adult pairings, since the former are often construed as a pseudolongitudinal design, providing information on the childhood correlates of adult outcomes (Burns, Moll, & Lauer, 1989; Ramirez, 1993; Ayatollahi, 1992; Price et al., 1990; Griffiths & Payne, 1976). With parent–offspring data alone, however, it is not possible to determine whether the observed similarity is due to shared genes, shared environment, or both. The same caveat applies to data on biological siblings reared together.

In contrast to the parent–offspring design, the adoption design compares adoptees to their adoptive and biological parents or adoptive and biological siblings. This comparison allows separation of familial similarity due to a shared environment from that due to shared genes. Both adult–child and adult–adult parent–offspring pairings have been used in adoption studies (Sorensen, Holst, & Stunkard, 1992; Sorensen, Price, Stunkard, & Schulsinger, 1989; Stunkard et al., 1986b; Price, Cadoret, Stunkard, & Troughton, 1987). As Sorensen et al. (1992) demonstrated, comparison of the similarity of these different parent–offspring pairings can provide insight into how the genetic control of obesity changes over time.

Although the collection of longitudinal data on obesity is more expensive and time-consuming than the collection of cross-sectional data, the benefits of longitudinal data cannot be overstated. As with twin data analyzed by birth cohort, genetically informative longitudinal data can address questions regarding age-related changes in heritability. But perhaps more importantly, by following a twin or adoptive cohort over time, one can not only determine the extent to which obesity tracks throughout the lifetime, but also partition this phenotypic tracking into genetic and environmental components. Consequently, we can ask whether individual differences in obesity created by the environment at a young age persist into adulthood and compare the persistence of these effects with that of inherited predispositions. We can also determine whether new genetic influences on obesity are expressed over the life span.

CROSS-SECTIONAL TWIN STUDIES

Overview

The classic twin design, employing MZ and DZ twins, has a number of attributes that make it appealing for investigating individual differences in obesity. Of primary importance is the power of the design. Since MZ twins share 100% of their genes, we expect them to correlate highly for any trait that is under strong genetic control. This expected high correlation provides the design with greater power for detecting genetic influences than a design that utilizes familial relationships with a lower genetic correlation. A second beneficial feature of the twin design is that it can detect and distinguish family environmental influences on obesity (such as food preferences) from genetic effects. Third, since the twin design inherently controls for age differences within the pair, twin similarity cannot be reduced by age-dependent genetic effects. These influences have the consequence of reducing the similarity of pairs of relatives who, because of their different ages, are not expressing the same genes. Fourth, as noted previously, when twin data are collected on restricted age cohorts, heritability analyses can reveal whether age-dependent changes in phenotypic means (e.g., increasing weight with increasing age) are accompanied by age-dependent changes in the relative effects of genes and environment on the phenotype. Finally, by obtaining separate heritability estimates from male–male and female–female twin pairs, the classic twin design can assess whether genes and environment have a differential impact on obesity across gender. When the studies of same-sex pairs are further extended to include opposite-sex twins, hypotheses regarding sex-limited expression of genetic effects can be tested (see Chapter 10).

Twin Studies of the Body Mass Index

These attributes of the twin design having been enumerated, it is not surprising to find that most of the data on individual differences in obesity across the life span have come from twin studies. With few exceptions, the majority of twin studies have utilized the body mass index (BMI) as a global measure of body size. The BMI is defined as weight (in kilograms) divided by the square of height (in meters). Although critics of this index will point out that it is an imperfect measure of fat mass or fat distribution, advocates of the BMI emphasize the ease of obtaining height and weight on a large number of related individuals and the moderately high correlation of BMI with indices of total body fat, percentage body fat, and subcutaneous fat deposits (Byard, Siervogel, & Roche, 1983; Selby et al., 1990). This pragmatic view of assessing obesity has provided us

TABLE 2. Heritability Estimates of the Body Mass Index Obtained from Cross-Sectional Twin Studies

Study	Number of pairs	Sex	Age	Heritability	Estimator	Comments
Wang, Ouyang, Wang, & Tang (1990)	75 MZ, 35 DZ	M and F same-sex	7–12	0.78	$2^* (r_{mz} - r_{dz})$	Data are age- and sex-adjusted.
Bodurtha et al. (1990)	148 MZ, 110 DZ	M and F same- and opposite-sex	11.1 ± 0.3	0.87	Maximum likelihood	No sex differences.
Korkeila et al. (1991)	2370 MZ, 4873 DZ	M and F same-sex	18–24 25–34 35–44 45–54	0.72 (M), 0.71 (F) 0.74 (M), 0.69 (F) 0.67 (M), 0.68 (F) 0.67 (M), 0.53 (F)	Maximum likelihood	Significant age and sex differences.
Stunkard et al. (1986a)	1974 MZ, 2097 DZ	M	20 ± 2.6 25-year follow-up	0.77 0.84	$2^* (r_{mz} - r_{dz})$	Heritability of BMI at follow-up is overestimated due to genetic dominance.

(continued)

TABLE 2. *(Continued)*

Study	Number of pairs	Sex	Age	Heritability	Estimator	Comments
Stunkard et al. (1990)[a]	93 MZA, 218 DZA, 154 MZT, 208 DZT	M and F same-sex	58.6 ± 13.6	0.74 (M), 0.69 (F)	Maximum likelihood	Genetic dominance for males.
Neale & Cardon (1992)—Virginia sample	2552 MZ, 3036 DZ	M and F same- and opposite-sex	18–90	0.69 (M), 0.75 (F)	Maximum likelihood	Age-adjusted. Genetic dominance for males.
Neale & Cardon (1992)—Australian sample	1703 MZ, 1866 DZ	M and F same- and opposite-sex	18–30 > 31	0.80 (M), 0.78 (F) 0.70 (M), 0.69 (F)	Maximum likelihood	Genetic dominance for young males and females.
Price & Gottesman (1991)[a]	34 MZA, 38 MZT	M and F same-sex	> 17	0.61 (MZA) 0.75 (MZT)	Direct estimates	Age- and sex-adjusted.
Hunt et al. (1989)	146 MZ, 162 DZ	M	34 ± 10	0.54	2* ($r_{mz} - r_{dz}$)	Age-adjusted.

[a](MZA, DZA) Twins reared apart; (MZT, DZT) twins reared together.

with a very good characterization of the nature of individual differences in the BMI.

Table 2 summarizes heritability estimates for the BMI obtained from several twin studies across the life span. Age and sex characteristics of the samples have been highlighted, as well as the method of obtaining the heritability estimate. The maximum likelihood model-fitting approach to twin data (Neale & Cardon, 1992) is currently the preferred method of analysis, since it allows one to jointly estimate the impact of additive genes, family environment (or, alternatively, genetic dominance or epistasis), and individual environment on a phenotype. Furthermore, the significance of each of these effects can be evaluated by removing it from the fully parameterized model and assessing the goodness of fit of the reduced model.

The most striking feature of Table 2 is the consistency in the heritability estimates. Despite methodological and sampling differences, the range of the estimates is narrow, with all but four estimates falling between 0.60 and 0.80. From those studies that have performed genetic analyses separately by birth cohorts [Korkeila et al., 1991; Neale & Cardon, 1992 (Australian sample)], there is an indication that the heritability of BMI declines over time in both males and females. This decline may reflect the developmental accumulation of environmental "insults," a hypothesis that can be directly tested with longitudinal data.

One apparent contradiction to this developmental decrease in heritability is found in the study by Stunkard, Foch, and Hrubec (1986a), which indicated that the heritability of the BMI in male veteran pairs increased from 0.77 at induction into the armed services (average age: 20) to 0.84, 25 years later. It should be noted, however, that the method used to estimate heritability in this analysis [2* 9$r$mz − $r$dz)] will inflate the estimate when the DZ twin correlation is less than one half the MZ correlation and genetic nonadditivity (dominance or epistasis) is indicated. If maximum-likelihood model-fitting methods were applied to these data, and nonadditive genetic effects were included in the parameterization, the broad-sense heritability estimate of the BMI would approximate the observed MZ correlation of 0.67, and the apparent increase in heritability over time would not be noted.

The study by Stunkard et al. (1986a) is not the only report of non-additive genetic influences on the BMI. In their analysis of older (mean age: 58) reared-apart and reared-together Swedish twins, Stunkard et al. (1990) found evidence for dominance effects on the BMI of male twins. Similarly, the analyses by Neale and Cardon (1992) of twin data from Virginia males and 18- to 30-year-old Australian males and females revealed significant nonadditive genetic variation. These findings are particularly noteworthy, since the twin design is not a very powerful method for detecting nonadditive genetic effects. Certainly, additional studies are required to replicate these findings—and to characterize more completely their age and sex dependence. If the importance of these effects is confirmed, then the question remains whether they can be explained by a major (dominant or recessive) gene for obesity. Some insight into this issue has already been gained through genetic segregation analyses, which are discussed at the end of this chapter.

Twin Studies of Fat Deposition and Patterning

Compared to twin studies on the BMI, there are relatively few twin studies that have assessed absolute amounts of fat and fat patterning. Recently, however, geneticists have been encouraged to study these additional indices of obesity given the confounding of the BMI with lean body mass and evidence for an independent relationship (after controlling for the BMI) between centralized, upper-body fat and cardiovascular risk factors (Donahue et al., 1987; Stevens, Gautman & Keil, 1993; Haffner, Mitchell, Hazuda, & Stern, 1991).

There are two child and two adult twin studies in the literature that have utilized skin folds and waist–hip circumference ratios to evaluate genetic influences on absolute and relative fat deposits. In general, skin folds have been used to evaluate the absolute amount of subcutaneous fat and differences in central vs. peripheral fat deposition. The waist–hip ratio is useful in quantifying the upper vs. lower body fat.

In their study of 222 pairs of same-sex twins of ages 3–15, Brook, Huntley, and Slack (1975) obtained age- and sex-corrected heritability estimates for triceps and subscapular skin folds of 0.46 and 0.98, respectively. When the twins were divided into two age cohorts (<10 and ≥ 10), significant differences were found in the heritability of triceps skin folds for both boys and girls, with genetic factors becoming markedly more important in the older children (heritability estimates for younger and older boys were 0.0 and 0.81, respectively, while those for younger and older girls were 0.56 and 0.90). The authors suggested that variable dietary factors may influence triceps skin folds in the younger children to a greater extent than in the older children. Bodurtha et al. (1990) also studied skin folds in 259 pairs of 11-year-old same-sex and opposite-sex twins. They concluded that the genetic influences on triceps skin folds were the same in both males and females and accounted for 75% of the phenotypic variance. Similar to the Brook et al. (1975) study, suprailiac thickness was somewhat more heritable in girls (0.87) than in boys (0.81), but there was no evidence for sex-specific genetic effects.

The interpretation of these heritability results for skin folds of children is limited by three factors. First, neither of the research groups attempted to correct the skin folds for overall body size (BMI) prior to their analysis. Given the high correlation of these measures with the BMI, it is not clear whether the heritability results reflect the heritability of the BMI or independent genetic influences on the skin folds. Second, the results do not reveal whether genetic influences on suprailiac skin folds are related to those on the triceps skin folds, i.e., whether there is a common genetic risk factor for an increase in subcutaneous fat. Multivariate genetic analyses of skin folds, along with the BMI, would address these first two issues (Neale & Cardon, 1992). Third, and perhaps most importantly, neither study explicitly demonstrated a relationship between juvenile and adult skin folds. For this reason, it is not known whether a genetic influence on *juvenile* skin folds is an important determinant of *adult* obesity and its health complications.

Additional twin data on fat distribution include those collected by Bouchard, Perusse, Leblanc, Tremblay, and Theriault (1988) as part of the Quebec Family Study. In this study, the authors fitted path-analytical models to data on adolescent twins (87 MZ pairs, age 17.0 ± 4.5; 69 DZ pairs, age 15.1 ± 4.1) as well as biological and adoptive parents and their offspring, first cousins, uncles/aunts–nieces/nephews, and biological siblings. From the analysis of the full data set, Bouchard et al. (1988) concluded that subcutaneous fat (as indexed by a sum of six skin folds) was minimally heritable (5%), while fat distribution (a ratio of extremity to trunk skin folds) was more highly heritable (30%). These results are provocative, since they suggest that *where* fat is deposited is more under genetic control than *how much* subcutaneous fat is deposited. If the twin data alone are examined, however, the sum of the six skin folds is slightly more heritable (89%) than fat distribution (80%), suggesting no real difference in the genetic control of absolute vs. relative fat deposits. It is possible that the discrepancy between the twin and family data reflects age-specific genetic influences or non-additive genetic effects, or both, on these phenotypes, but the specific analytical method used in this investigation was not designed to address these issues. Another limitation of this study was that the influence of body size (BMI) on the measures of total fat and fat distribution was not accounted for in the analysis. For this reason, it is not known whether the genetic influences on subcutaneous fat and fat distribution are independent of those on the BMI.

A final twin study of fat distribution has been reported by Selby et al. (1990). They examined the heritability of subscapular and triceps skin folds, their ratio, and the waist–hip ratio in a cohort of 265 pairs of male twins of ages 59–70 who were members of the National Academy of Sciences–National Research Council (NAS-NRC) Twin Registry. The authors adjusted each of these variables for the BMI prior to estimating heritability. The adjustment was more important for the raw skin folds (correlating 0.53–0.62 with the BMI) than for the ratios (correlating 0.17–0.47 with the BMI). Genetic factors, uncorrelated with the BMI, were found to explain a large amount of the variance of subscapular and triceps skin folds (63% and 83%, respectively) and a smaller amount of variance in the subscapular–triceps ratio (24%) and waist–hip ratio (31%)—results contrary to those of the Quebec Family Study. The authors suggested that behavioral variables (including smoking and exercise) may be a more important determinant of fat distribution than direct genetic influences.

Summary

Using twin data, investigators have made progress in assessing the nature of individual differences in indices of obesity. Results for the BMI are consistent across studies: Heritability is moderately high (ranging from 0.60 to 0.80), environmental influences appear to become slightly more important in older age, and nonadditive genetic effects (i.e., genetic dominance) may influence the phenotype. Compared to the BMI, there is not as much information available on absolute amounts of fat and fat patterning in twins. Although skin folds and waist–hip ratios appear to have a genetic basis, additional studies are required to assess the specificity of genetic influences on measures of fat deposition and differences in these effects across age and gender.

STUDIES OF BIOLOGICAL PARENTS AND THEIR OFFSPRING

Overview

One strategy to reduce the adverse health effects of adult obesity is to identify developmental precursors of obesity and then attempt to alter the precursors prior to the onset of complications. In the absence of longitudinal data, parent–offspring data provide useful information on the developmental precursors of adult obesity and its health complications. In the past few years, several different studies have been published that use the parent–offspring relationship to explore issues in prediction. These studies have not addressed parent–offspring similarity for global measures of body size per se, but rather have focused on intergenerational similarity for fat distribution, the relationship between early-onset obesity and family risk for obesity, and the relationship between child obesity and adult cardiovascular mortality. It is on these newer applications of the parent–offspring design that the following discussion focuses.

Parent–Offspring Similarity for Indices of Fat Distribution

A study by Donahue, Prineas, Gomez, and Hong (1992) evaluated parent-offspring similarity for the waist–hip ratio after controlling for parent and child BMI and child height. They analyzed data on 712 families, with children of ages 13–17, using multiple regression. Their results revealed significant relationships between all parent–offspring pairs (father–son, mother–son, father–daughter, and mother–daughter), with the effect being greatest for father–daughter pairs and smallest for father–son pairs.

In a smaller study of 194 families with children ages 11–20, Ramirez (1993) presented parent–offspring intraclass correlations for the waist–hip ratio adjusted for BMI and age. Similarly to the Donahue et al. (1992) study, Ramirez concluded that fathers and daughters were most similar for adjusted waist–hip ratio, with a significant intraclass correlation of 0.21. The mother–daughter correlation (0.17) also reached significance, while the mother–son (0.07) and father–son (0.14) correlations were non-significant. These latter findings could reflect the small sample size of the studies, rather than the absence of parent–offspring similarity.

In general, the results from these two studies indicate not only that relative

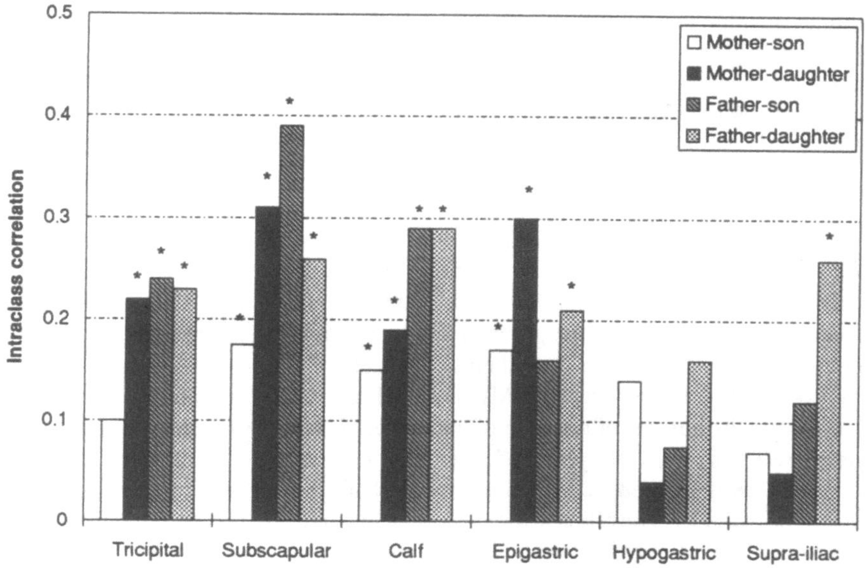

FIGURE 1. Parent–offspring correlations for six sites of subcutaneous fat, measured through ultrasound imaging. Data from Ramirez (1993). $*p < 0.05$.

fat distribution is independent of body size, but also that factors that uniquely influence fat patterning are correlated across generations. Consequently, predictions about parental risk for cardiovascular complications can be made by assessing waist–hip ratios in children. Furthermore, if the parent–offspring similarity is mediated by genetic effects or shared environmental influences that do not change over time, then we would expect a child with a high waist–hip ratio to be at an increased risk for cardiovascular complications in older age. It must be emphasized, however, that the parent–offspring correlations are small to moderate, and if they are due only to genetic effects, their magnitude would imply a cross-generational heritability of 0.14–0.40. Certainly, environmental or genetic influences unique to each generation also play a large role in determining the relative distribution of fat.

Ramirez (1993) also presented parent–offspring correlations for subcutaneous adipose tissue thickness, as determined from an ultrasound scanner. This report was the first to use this technology in a genetic design. We have summarized the correlations (corrected for age and BMI) in Fig. 1. The values range from 0.03 (for the mother–daughter hypogastric correlation) to 0.39 (for the father–son subscapular correlation). For the six sites examined, the general finding was of a lesser parent–offspring similarity for those sites on the lower abdomen (hypogastric and suprailiac) than for sites elsewhere (tricipital, subscapular, calf, and epigastric). The authors suggested that deposits on these sites were more easily modified with diet and exercise. When the similarity of different parent–offspring types was compared, the mother–son correlation was commonly smaller than the other pairings; however, tests of heterogeneity across pairings indicate that none of the observed differences was significant. Larger data sets, however, may reveal significant gender-related trends that would have implications for child–parent predictions.

Early-Onset Obesity and Familial Risk

Two additional studies of the onset of obesity in childhood and family risk for obesity and mortality deserve mention. In a multicenter study involving 1743 first-degree adult relatives of 566 obese patients, Price et al. (1990) assessed the age of onset of obesity in the proband and its relationship to the number of first-degree adult relatives who were themselves overweight or obese (as determined by body silhouette reports or BMI data). Their results indicated a significant increase in the number of obese family members for those probands whose age of onset was less than 10 years of age. The relative risk for obesity in these relatives, as compared to those with the proband having age of onset after 10 years, was 2.14 (controlling for age, sex, and sample). The authors suggested that age of onset may be used to refine future genetic analyses of obesity as a marker of greater genetic or familial risk. It will also perhaps help define more etiologically homogeneous groups of families for the purpose of genetic segregation or linkage analysis (see Chapter 13 for a discussion of such methods).

In an additional study of children and their adult relatives, Burns et al. (1992) adopted a simple and direct approach for assessing the relationship between childhood obesity and adult cardiovascular complications. They collected total and cause-specific mortality data on 387 first- and second-degree relatives of children participating in the Muscatine Ponderosity Family Study. The children (initially ages 6–12) represented two weight classes (persistently lean or persistently heavy, as determined from the quintiles of age- and sex-specific weight distributions, over 4 years of assessment) and a control group. Using a Cox proportional hazard model, Burns and colleagues compared the risk of relatives' death from cardiovascular disease in each of the lean and heavy groups against the control group. While being lean did not significantly decrease risk for cardiovascular death in the relatives, being heavy significantly increased risk (relative risk = 1.41, c.i. 1.02–1.98). Moreover, when the heavy group was divided into those children who were hypertensive and non-hypertensive, the effect was even greater for the hypertensive children (relative risk = 2.20, c.i. 1.43–3.37). The authors suggested that persistent obesity in children may be a marker for persistent obesity in adult relatives, which itself is related to more proximal cardiovascular risk factors including increased blood pressure and apoprotein B and triglyceride levels.

Summary

Although parent–offspring studies are not the best design for assessing the magnitude of genetic influences on obesity, they can be used to identify childhood risk factors of adult obesity and its health complications. The results from the studies presented above indicate that a child's waist–hip ratio, amount of subcutaneous fat, age of onset of obesity, and BMI are related to, respectively, his or her parent's waist-hip ratio, amount of subcutaneous fat, obesity, and mortality. In order to conclude from these data that the children themselves are at greater risk for the adult outcomes, one must make the assumption that the influences responsible for child–adult similarity remain stable over time. If, for example, the observed similarity reflects a transient environmental influence that parents and children share, predictions about the child's adult outcome are not valid.

ADOPTION STUDIES

Overview

As mentioned previously, adoption studies (which compare the similarity of adoptees to their adoptive and biological parents or siblings) are usually preferred over parent–offspring studies because they allow the separation of familial influences that are environmental in origin (e.g., shared diet) from those that are genetic. Adoption studies, however, are not without their own limitations. Issues regarding selective placement and nonrepresentation of the population can be of concern in adoption studies of some phenotypes. However, when these potential biases have been evaluated in adoption studies of obesity, they have not been detected (Stunkard et al., 1986b; Price et al., 1987).

Danish Adoption Studies

The first large adoption study that analyzed BMI data on adult adoptees (mean age: 42.2 ± 8.1 years) and their biological and adoptive parents was reported by Stunkard et al. (1986b). To determine the extent of parent–offspring similarity for obesity, they first selected 540 adoptees from the Danish adoption registry to represent four weight classes (I, thin; II, median; III, overweight; IV, obese). These weight classes were determined from percentiles of age-specific and sex-specific BMI distributions. The authors then computed the mean BMI for the biological and adoptive parents of these adoptees by their weight class. Their results, summarized in Figure 2, are presented separately by gender of the biological and adoptive parents. The relationships between the weight class of the adoptees and the mean BMI of both biological mother and father were significant. When the data were analyzed separately by the gender of the children, mother–daughter similarity was significant, but mother–son, father–daughter, and father–son similarity were not. In contrast to the biological parent–offspring trends, none of the adoptive parent–offspring trends was significant.

Sorensen et al. (1989) extended the Danish adoption study of obesity to include data on 210 full siblings and 1358 maternal and paternal half-siblings of the adoptees. The mean BMIs for these individuals, by weight class of adoptee, are also given in Fig. 2. Similarly to the parental data, there is a marked increase

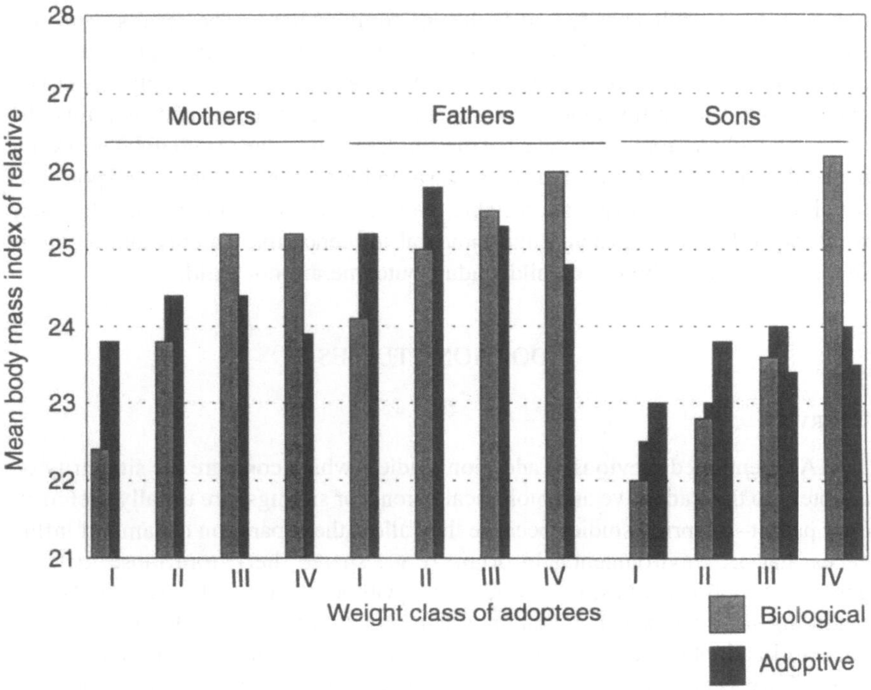

FIGURE 2. Mean BMI of biological and adoptive relatives of Danish adult adoptees, divided into four weight classes: (I) thin; (II) median; (III) overweight; (IV) obese. Biological siblings are contrasted in this figure with maternal and paternal half-siblings. The increasing trends for the mean BMI of biological relatives are significant, while those for the adoptive relatives are not. Data from Stunkard et al. (1986b) and Sorensen et al. (1989).

in the BMI of the full siblings with increasing weight class of the adoptees. When a multiple regression analysis, controlling for age and sex, was applied to these data, the increasing trend was highly significant ($p < 0.0001$). In contrast, when the data from the maternal and paternal half-siblings were analyzed, there was only a moderate, though significant ($p < 0.02$), increasing trend in the BMI of the half-siblings. There was not a significant difference in the trend for the maternal vs. the paternal half-siblings, although separate analyses of the maternal and paternal half-sibling data showed that the linear increase was significant only for the maternal half-siblings. The authors noted that the positive regression coefficients for both full and half-siblings, and the difference in their magnitude, provided supportive evidence for a genetic influence on the BMI.

Iowa Adoption Study

An additional adoption study of 357 adult Iowa adoptees and their biological and adoptive parents was described by Price et al. (1987). BMI data were obtained on the adoptees when they were of ages 18–38, their biological parents at the time of adoption, and their adoptive parents at an average age of 54. Initial correlational results indicated that only the similarity between the BMI of biological mothers and daughters ($r = 0.4$) reached statistical significance. The remaining biological correlations were all positive, but nonsignificant (0.18, father–daughter; 0.15, mother–son; 0.08, father–son). The adoptive parent–offspring correlations, which directly assess the impact of the home-rearing environment on the adoptees' BMI, ranged from −0.09 to 0.09 and were individually nonsignificant.

Price et al. (1987) applied a hierarchical multiple regression model to these data to jointly evaluate genetic similarity for BMI and test for the influence of measured environmental variables—including rural residence and a disturbed rearing environment (psychopathology or divorce in the adoptive parents)—on the adoptees' BMI. The results of the regression analysis indicated that the measured environmental effects did play a significant (but small) role in influencing the BMI of the adoptees. The influence of the biological mother's BMI on the adoptee's BMI was highly significant, while the effect of the biological father's BMI, while positive, was nonsignificant. It is of interest to note that this increased mother–child similarity was found in the Danish data as well and may indicate a maternal effect on adult BMI.

Summary

Results from the Danish and Iowa adult adoption studies support the results from twin studies that show that genes play a significant role in the etiology of the BMI. There is, however, an indication in the adoption data of Price et al. (1987) that features of the family environment (including rural residence and disruptive home) can also influence obesity. Additional data are needed to determine whether these environmental effects are important in other populations. Finally, there is suggestive evidence from both studies that maternal effects may increase biological parent–offspring and half-sibling similarity.

LONGITUDINAL STUDIES

Overview

In the discussion of cross-sectional studies, we drew attention to the fact that genetic influences and, to a much lesser extent, environmental influences on obesity are correlated between family members, across generations. It is these genetic and environmental influences that create similarity between biological or adoptive parents and their offspring. What is not known from the parent–offspring data is whether there are *familial* influences on obesity *unique to each generation*. These influences may be of three types: (1) age-specific, additive genetic effects; (2) genetic dominance effects, which correlate zero between parents and their offspring; and (3) family environmental effects (such as a shared diet or exercise routine) that operate on one generation but not the other. An increase in any one of these effects can diminish parent–offspring similarity. As a consequence, one may make the erroneous conclusion that familial factors are not important determinants of obesity, when it is only the case that they are not important determinants of cross-generational similarity.

This rather subtle point about age-dependent effects can have quite important consequences for the interpretation of genetic data, since it tells us that a high heritability of a phenotype at one age implies nothing about the stability of gene expression over time. In much the same way, a *low heritability* at one age gives no information about the stability of environmental influences, unique to the individual, over time. Large individual environmental influences on a phenotype assessed at one age may be uncorrelated, or perfectly correlated, with environmental influences at a later age. Just as parent–offspring data alone provide no information on age-specific family influences, they provide no information on these age-to-age, environmental correlations.

Some insight into the issue of age-dependent familial influences can be gained by comparing cross-sectional twin data from adults and juveniles to adult–juvenile, parent–offspring data. If familial influences on the phenotypes of the twins are of greater magnitude than those that create parent–offspring similarity, age-specific family effects are implied. One could conclude from this type of analysis that genes and family environment are important determinants of obesity at a given age, but are not responsible for age-to-age correlations. What cannot be gained from such an analysis is whether factors that are *not* familial in origin create age-to-age continuity. The importance of individual environmental continuity can be assessed only through genetically informative, longitudinal data. It is these data that allow one to determine whether genes or the environment or both create phenotypic continuities and discontinuities over time.

Empirical Evidence of Tracking

Prior to conducting a genetically informative, longitudinal study of obesity, it is useful to know the extent to which the phenotype is expected to correlate (or track) over an individual's lifetime. Data from the British National Survey of Health and Development (Stark, Atkins, Wolff, & Douglas, 1981) provided an excellent characterization of the tracking of relative weight in a cohort of 5362 children at ages 6, 7, 11, 14, and 26. Their correlational data are presented in Figure 3, for males and females separately. Of note are the moderate to large correlations (ranging from 0.58 to 0.78) between any two consecutive assessments and the decrease between these correlations as the time between assessment periods increases. The implications of such a pattern of correlations are discussed, in the context of blood pressure, in Chapter 6. The pattern is similar for males and females, although the female correlations are slightly higher.

Other longitudinal studies of relative weight or obesity throughout child-

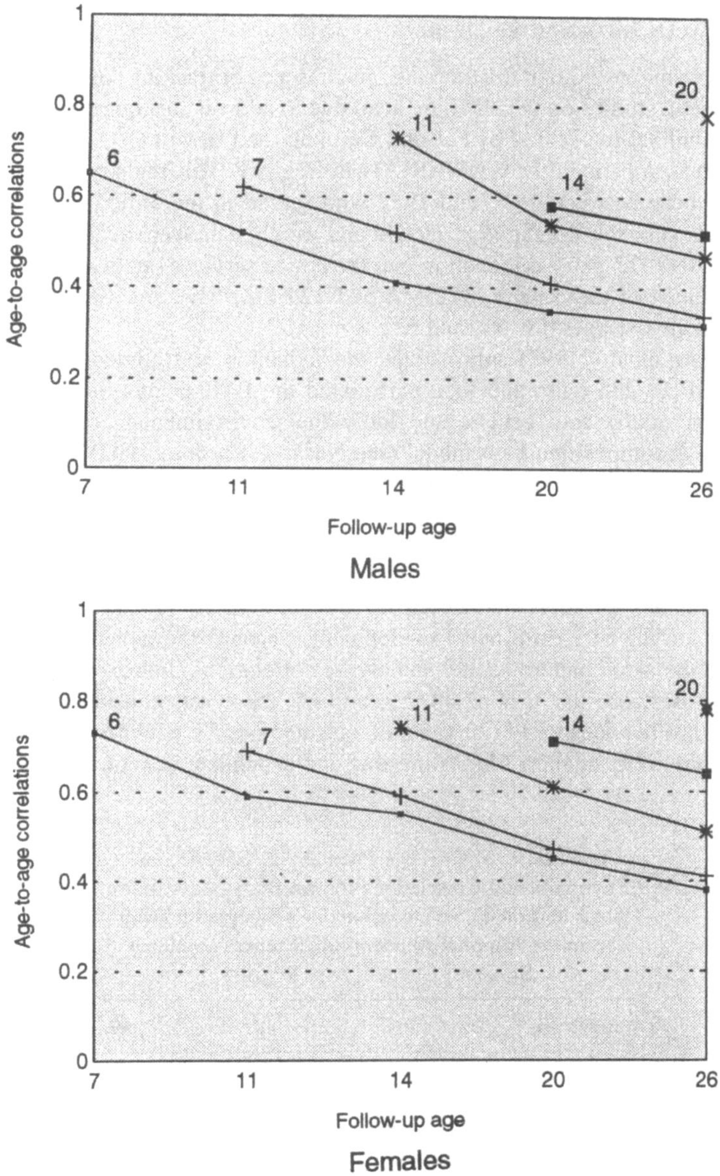

FIGURE 3. Age-to-age correlations for relative weight reported by Stark et al. (1981).

hood and into early adulthood have reached similar conclusions about continuities over this time period (Berkowitz, Agras, Korner, Kraemer, & Zeanah, 1985; Charney, Goodman, McBridge, Lyon, & Pratt, 1976; Melbin & Vuille, 1976). Less is known about the childhood correlates of obesity in middle and older age, owing to the difficulty of obtaining such data.

Genetically Informative Designs

How can one explain the tracking, and absence of tracking, of obesity? Two large genetic studies on the BMI are available to address this question. The first is a longitudinal twin study by Fabsitz, Carmelli, and Hewitt (1992) that utilized data from twin pairs in the NAS-NRC Twin Registry. This registry is comprised of males born between 1917 and 1927 who served in the armed forces during World War II or the Korean War. Height and weight data were available on 1974 MZ and 2097 DZ pairs at induction into the armed services (at an average age of 20). A subset of these pairs, 121 MZ and 124 DZ, have since participated in clinical exams at ages 48, 59, and 65.

Fabsitz et al. (1992) summarized the induction and clinical examination data for these 245 pairs and then partitioned the BMI phenotypic, age-to-age correlation matrix into genetic and individual environmental components. A Choleksy decomposition path model (see Neale & Cardon, 1992) was used for the analysis. The correlations obtained from their results are shown in Table 3. There were three noteworthy findings. First, phenotypic continuity was moderate between early and middle adulthood, but quite high within middle adulthoood. This pattern of increasing correlations with decreasing time between measurement occasions parallels that described by Stark et al. (1981) in the British National Survey of Health and Development. Second, the genetic correlations followed the same pattern at the phenotypic correlations, but exceeded them in all cases. Between the ages of 48, 59, and 65, there was considerable genetic stability, but between early (20 years of age) and middle adulthood, the genetic correlations were smaller, suggesting that age-dependent genetic influences are

TABLE 3. Age-to-Age Phenotypic, Genetic, and Environmental Correlations for the Body Mass Index Measured on Three Occasions in a Sample of Twins from the National Academy of Sciences–National Research Council Twin Registry[a]

Average age	20	48	59
48	**0.48**	—	—
	0.64	—	—
	0.21	—	—
59	**0.42**	**0.85**	—
	0.73	*0.98*	—
	0.02	0.48	—
65	**0.39**	**0.72**	**0.90**
	0.64	*0.96*	*0.93*
	0.02	0.43	0.36

[a] From Fabsitz et al. (1992). **Phenotypic** correlations are shown in **boldface** type, *genetic* correlations in *italic* type, and environmental correlations in plain type.

operating on the phenotype. Finally, the pattern of environmental correlations indicated moderate stability of the individual influences on the BMI at ages 48, 59, and 65, but much smaller correlations between the environmental effects that create individual differences at age 20 and those that influence BMI in middle age. The authors concluded from this analysis that genes are primarily responsible for age-to-age continuity, with environmental effects playing a minor role in middle age. There was also evidence that an independent genetic influence on BMI appears during early middle age, increasing the total variance of BMI and persisting into latter middle age.

Cardon (1993) (see also Chapter Seven) has also described a longitudinal analysis of the BMI in juvenile adoptive and biological siblings and their parents, who were participating in the Colorado Adoption Project. This study included 245 families, and height and weight data were collected on the children on as many as 9 occasions, from birth to age 9. In contrast to the results of Fabsitz et al. (1992) for adults, Cardon found no evidence for age-to-age continuities in individual environmental effects on the BMI in early childhood. Instead, these large effects, accounting for as much as 68% of the phenotypic variance, were occasion-specific. The age-to-age phenotypic correlations, which grew smaller as the time between measurement occasions increased, were solely explained by genetic influences. In addition, there was evidence for specific genetic influences operating at each age of assessment.

Summary

Overall, the results from these longitudinal analyses indicate that genes create both continuity and change in the BMI from birth to age 65. There appear to be common and age-specific genetic influences on the BMI over the life span. In contrast, environmental effects do not explain BMI tracking during early childhood, but they become more important in middle age. Consequently, dietary or exercise modifications in early childhood may have little impact on long-term changes in adiposity.

CONCLUSIONS, LIMITATIONS, AND FUTURE DIRECTIONS

Twin, adoption, and family studies of indices of obesity have reached a similar conclusion: Genes influence adiposity. We are just beginning to consider that different sets of genes may be responsible for individual differences in obesity over time, across gender, and across methods of assessment. It is only with additional longitudinal and multivariate genetic studies, designed to identify age-specific and phenotypic-specific genetic influences, that further insight into these issues will be gained.

The existence of age-specific genetic influences has important implications for studies aimed at identifying major genes that influence obesity, through genetic segregation and linkage analyses. Already, three large segregation analyses of the BMI (Moll, Burns, & Lauer, 1991; Price, Ness, & Laskarzewski, 1989; Province, Arnqvist, Keller, Higgins, & Rao, 1990) have found support for a single recessive major gene effect on the phenotype, while two have not (Karlin, Williams, Jensen, & Farquhar, 1981; Zonta, Jayakar, Bosisio, Galante, & Pennetti, 1987). Borecki, Rice, Bouchard, and Rao (1993) raised the possibility that the apparent contradictory results reflect developmental change in gene expression. When they modeled these effects in their analysis of the BMI from the Quebec Family Study, they found evidence for a major gene effect on the phenotype that was not detected through traditional segregation analysis methods.

Different indices of obesity may also influence the results of segregation and linkage analyses. The greatest success in this area will most likely be obtained by selecting phenotypes that are not, a priori, known to be heterogeneous. The BMI, for example, is a function of both fat mass and fat-free mass. Segregation and linkage studies of this variable may be identifying genes that are responsible for both components, rather than those that influence fat mass alone. Applying segregation and linkage analysis to direct measures of fat mass (see Rice, Borecki, Bouchard, & Rao, 1993) may prove to be a better strategy for findings genes responsible for obesity.

An additional factor that will complicate the search for genes that influence obesity is the existence of gene–environment interactions. These influences have been shown to be important by Bouchard et al. (1990) in their feeding study of 12 pairs of adult MZ male twins. Even though they were given the same caloric excess for 84 days, and followed a similar sedentary life-style, the males varied in the amount of weight gained. However, there was a significant intrapair similarity in weight change, suggesting that the genotypes that were represented in the study responded differently to an environmental challenge. It is not known whether these interactions are important in females or at younger and older ages.

From this discussion, it appears that we have only scratched the surface in understanding the genetic and environmental etiology of obesity. The challenge of future studies will be to identify *which* genes influence *which* aspects of obesity, how genes interact with environmental effects, and how gene expression changes over time. With this information, we can hope to design more effective intervention strategies aimed at reducing obesity-related morbidity and mortality.

REFERENCES

Ayatollahi, S. M. T. (1992). Obesity in school children and their parents in southern Iran. *International Journal of Obesity, 16,* 845–850.

Barrett-Connor, E. L. (1985). Obesity, atherosclerosis, and coronary artery disease. *Annals of Internal Medicine, 103,* 1010–1019.

Baumgartner, R. N., Roche, A. F., Chumlea, W., Siervogel, R. M., & Glueck, C. J. (1987). Fatness and fat patterns: Associations with plasma lipids and blood pressures in adults, 18 to 57 years of age. *American Journal of Epidemiology, 126,* 614-628.

Berkowitz, R. I., Agras, W. S., Korner, A. F., Kraemer, H. C., & Zeanah, C. H. (1985). Physical activity and adiposity: A longitudinal study from birth to childhood. *Journal of Pediatrics, 106,* 734–738.

Blair, D., Habicht, J. P., Sims, E. A. H., Sylvester, D., & Abraham, S. (1984). Evidence for an increased risk for hypertension with centrally located body fat and the effect of race and sex on this risk. *American Journal of Epidemiology, 119,* 526–540.

Bodurtha, J. N., Mosteller, M., Hewitt, J. K., Nance, W. E., Eaves, L. J., Moskowitz, W. B., Katz, S., & Schieken, R. M. (1990). Genetic analysis of anthropometric measures in 11-year-old twins: The Medical College of Virginia Twin Study. *Pediatric Research, 28,* 1–4.

Borecki, I. B., Bonney, G. E., Rice, T., Bouchard, C., & Rao, D.C. (1993). Influence of genotype-dependent effects of covariates on the outcome of segregation analysis of the body mass index. *American Journal of Human Genetics, 53,* 676–687.

Bouchard, C., Perusse, L., Leblanc, C., Tremblay, A., & Theriault, G. (1988). Inheritance of the amount and distribution of human body fat. *International Journal of Obesity, 12,* 205–215.

Bouchard, C., Tremblay, A., Despres, J., Nadeau, A., Lupien, P. J., Theriault, G., Dussault, J., Moorjani, S., Pinault, S., & Fournier, G. (1990). The response to long-term overfeeding in identical twins. *New England Journal of Medicine, 322(32),* 1477–1482.

Bray, G. A. (1985). Complications of obesity. *Annals of Internal Medicine, 103,* 1052–1062.

Brook, G. D., Huntley, R. M. C., & Slack, J. (1975). Influence of heredity and environment in the determination of skinfold thickness in children. *British Medical Journal, 2,* 719–721.

Burns, T. L., Moll, P. P., & Lauer, R. M. (1989). The relation between ponderosity and coronary risk factors in children and their relatives: The Muscatine Ponderosity Family Study. *American Journal of Epidemiology, 129,* 973–987.

Burns, T. L., Moll, P. P., & Lauer, R. M. (1992). Increased familial cardiovascular mortality in older schoolchildren. The Muscatine Ponderosity Family Study. *Pediatrics, 89*(2), 262–268.

Byard, P. J., Siervogel, R. M., & Roche, A. F. (1983). Sibling correlations for weight/stature^2 and calf circumference: Age changes and possible sex linkage. *Human Biology, 55,* 677–685.

Cardon, L. R. (1983). Height, weight, and obesity. In R. Plomin, J. C. DeFries, & D. W. Fulker (Eds.), *Nature and nurture druing middle childhood* (pp. 165–172). London: Blackwell.

Charney, E., Goodman, H. C., McBridge, M., Lyon, B., & Pratt, R. (1976): Childhood antecedents of adult obesity: Do chubby infants become obese adults? *New England Journal of Medicine, 295,* 6–9.

Donahue, R., Abbot, P., Bloom, R., Reed, E., & Katsuhiko-Yano, D. M. (1987). Central obesity and coronary heart disease in men. *Lancet, 1(8537),* 821–824.

Donahue, R. P., Prineas, R. J., Gomez, O., & Hong, C. P. (1992). Familial resemblance of body fat distribution: The Minneapolis Children's Blood Pressure Study. *International Journal of Obesity, 16,* 161–167.

Fabsitz, R. R., Carmelli, D., & Hewitt, J. K. (1992). Evidence for independent genetic influences on obesity in middle age. *International Journal of Obesity and Related Metabolic Disorders, 16,* 657–666.

Griffiths, M., & Payne, P. R. (1976). Energy expenditure in small children of obese and non-obese parents. *Nature, 260,* 698–700.

Haffner, S. M., Mitchell, B. D., Hazuda, H. P., & Stern, M. P. (1991). Greater influence of central distribution of adipose tissue on incidence of non-insulin dependent diabetes in women than men. *American Journal of Clinical Nutrition, 53,* 1312-1317.

Hunt, S. C., Hasstedt, S. J., Kuida, H., Stults, B. M., Hopkins, P. N., & Williams, R. R. (1989). Genetic heritability and common environmental components of resting and stressed blood pressures, lipids, and body mass index in Utah pedigrees and twins. *American Journal of Epidemiology, 129,* 625–638.

Karlin, S., Williams, P. T., Jensen, S., & Farquhar, J. W. (1981). Genetic analysis of the Stanford LRC family study data. I. Structural exploratory data analysis of height and weight measurements. *American Journal of Epidemiology, 113,* 307–324.

Korkeila, M., Kaprio, J., Rissanen, A., & Koskenvuo, M. (1991). Effects of gender and age on the heritability of the body mass index. *International Journal of Obesity, 15,* 647–654.

Melbin, T., & Vuille, J.-C. (1976). Weight gain in infancy and physical development between 7 and 10½ years of age. *British Journal of Preventive and Social Medicine, 30,* 233–238.

Moll, P. P., Burns, T. L., & Lauer, R. M. (1991). The genetic and environmental sources of body mass index variability: The Muscatine Ponderosity Family Study. *American Journal of Human Genetics, 49,* 1243–1255.

Neale, M. C., & Cardon, L. R. (1992). *Methodology for genetic studies of twins and families.* Dordrecht, The Netherlands: Kluwer Academic Press.

Price, R. A., Cadoret, R. J., Stunkard, A. J., & Troughton, E. (1987). Genetic contributions to human fatness: An adoption study. *American Journal of Psychiatry, 144,* 1003–1008.

Price, R. A., & Gottesman, I. I. (1991). Body fat in identical twins reared apart: Roles for genes and environment. *Behavior Genetics, 21,* 1–6.

Price, R. A., Ness, R., & Laskarzewski, P. (1989). Common major gene inheritance of extreme overweight. *Human Biology, 62,* 747–765.

Price, R. A., Stunkard, A. J., Ness, R., Wadden, T., Heshka, S., Kanders, B., & Cormillot, A. (1990). Childhood onset (age <10) obesity has high familial risk. *International Journal of Obesity, 14,* 185–195.

Province, M. A., Arnqvist, P., Keller, J., Higgins, M,. & Rao, D. C. (1990). Strong evidence for a major gene for obesity in the large, unselected, total Community Health Study of Tecumseh. *American Journal of Human Genetics, 47(Supplement),* A143.

Ramirez, M. E. (1993). Familial aggregation of subcutaneous fat deposits and the peripheral fat distribution pattern. *International Journal of Obesity and Related Metabolic Disorders, 17,* 63–68.

Rice, T., Borecki, I. B., Bouchard, C., & Rao, D. C. (1993). Segregation analysis of fat mass and other body composition measures derived from underwater weighing. *American Journal of Human Genetics, 52,* 967–973.

Selby, J. V., Newman, B., Quesenberry, C. P., Fabsitz, R. R., Carmelli, D., Meaney, F. J., & Selmenda, C. (1990). Genetic and behavioral influences on body fat distribution. *International Journal of Obesity, 14,* 593–602.

Sorensen, T. I. A., Holst, C., & Stunkard, A. J. (1992). Childhood body mass index—genetic and familial environmental influences assessed in a longitudinal adoption study. *International Journal of Obesity, 16,* 705–715.

Sorensen, T. I. A., Price, R. A., Stunkard, A. J., & Schulsinger, F. (1989). Genetics of obesity in adult adoptees and their biological siblings. *British Medical Journal, 298,* 87–90.

Stark, O., Atkins, E., Wolff, O. H., & Douglas, J. W. B. (1981). Longitudinal study of obesity in the National Survey of Health and Development. *British Medical Journal, 283,* 13–17.

Stevens, J., Gautman, S. P., & Keil, J. K. (1993). Body mass index and fat patterning as correlates of lipids and hypertension in an elderly, biracial population. *Journal of Gerontology, 48,* M249–254.

Stunkard, A. J., Foch, T. T., & Hrubec, Z. (1986a). A twin study of human obesity. *Journal of the American Medical Association, 256,* 51–54.

Stunkard, A. J., Harris, J. R., Pedersen, N. L., & McClearn, G. E. (1990). The body-mass index of twins who have been reared apart. *New England Journal of Medicine, 322,* 1483–1487.

Stunkard, A. J., Sorensen, T. I. A., Hanis, C., Teasdale, T. W., Chakraborty, R., Schull, W. J., & Schulsinger, F. (1986b). An adoption study of human obesity. *New England Journal of Medicine, 314,* 193–198.

Van Itallie, T. B. (1985). Health implications of overweight and obesity in the United States. *Annals of Internal Medicine, 103,* 983–988.

Wang, Z., Ouyang, Z., Wang, D., & Tang, X. (1990). Heritability of blood pressure in 7- to 12-year-old Chinese twins, with special reference to body size effects. *Genetic Epidemiology, 7,* 447–452.

Zonta, L. A., Jayakar, S. D., Bosisio, M., Galante, A., & Pennetti, V. (1987). Genetic analysis of human obesity in an Italian sample. *Human Heredity, 37,* 129–139.

CHAPTER NINE

Genetics and Eating Disorders

JEANINE R. SPELT AND JOANNE M. MEYER

INTRODUCTION

Anorexia nervosa and bulimia nervosa are two of the more physiologically severe psychiatric disorders in adolescents and young adults, yet our understanding of their causes remains limited. Symptoms of these disorders have been documented throughout history. Cases similar to what we now call anorexia were described as early as 700 A.D., when a Portuguese princess was said to have starved herself for so long that she began to look like a man (Lacey, 1982). The distinction between bulimia and anorexia emerged in the late 1970s, and the term "bulimia" was first applied to binge eating and purging behaviors by G. F. M. Russell in 1979 (Halmi, 1987). Diagnostic criteria for anorexia and bulimia nervosa appeared in the third edition of the *Diagnostic and Statistical Manual of Mental Disorders,* American Psychiatric Association (APA, 1980).

Although the eating disorders have been organized into functional diagnostic categories, etiological explanations continue to be investigated. Current research attempts to explain the onset of both disorders with developmental models that structure the effects of hypothesized biological, psychological, and social etiological factors. Genetic effects are not often explicitly included in these developmental models, yet their importance in the eventual onset of eating disorders has been inferred from twin and family studies.

In this chapter, we begin with a description of the diagnostic criteria for anorexia nervosa and bulimia nervosa and then review the epidemiological trends

JEANINE R. SPELT • Department of Clinical Psychology, Virginia Commonwealth University, Richmond, Virginia 23284. JOANNE M. MEYER • Department of Human Genetics, Medical College of Virginia, Virginia Commonwealth University, Richmond, Virginia 23298.

Behavior Genetic Approaches in Behavioral Medicine, edited by J. Rick Turner, Lon R. Cardon, and John K. Hewitt. Plenum Press, New York, 1995.

in expression of these disorders. Biopsychosocial models for the development of eating disorders are illustrated, followed by a review of the evidence supporting genetic influences on individual susceptibility to anorexia and bulimia nervosa.

CASE DEFINITION

Two major controversies have arisen in developing diagnostic criteria for the eating disorders. The first is a debate between the concept of a continuum of eating behaviors and a belief that anorexia and bulimia are distinct disorders unrelated to normal dieting. Nylander (1971) was the first to discuss the possibility of a continuum of eating behaviors, suggesting that extreme dieting might precipitate the onset of more severe restricting or purging behaviors. In a review of the recent research in this area, Hsu (1989) also argued in favor of the continuum hypothesis. Polivy and Herman (1987), on the other hand, have considered a series of studies designed to address the continuity issue and arrived at a different conclusion. Their review did support the existence of behaviors that

TABLE 1. DSM-III-R and DSM-IV Diagnostic Criteria for the Eating Disorders[a]

DSM-III-R	DSM-IV
Anorexia nervosa	
A. Refusal to maintain body weight over a minimal normal weight for age and height, e.g., weight loss leading to maintenance of body weight 15% below that expected; or failure to make expected weight gain during period of growth, leading to body weight 15% below that expected.	A. Refusal to maintain body weight at or above a minimally normal weight for age and height (e.g., weight loss leading to maintenance of body weight less than 85% of that expected; or failure to make expected weight gain during period of growth, leading to body weight less than 85% of that expected).
B. Intense fear of gaining weight or becoming fat, even though underweight.	B. Intense fear of gaining weight or becoming fat, even though underweight.
C. Disturbance in the way one's body weight, size or shape is experienced, e.g., the person claims to "feel fat" even when emaciated, believes that one area of the body is "too fat" even when obviously underweight.	C. Disturbance in the way in which one's body weight or shape is experienced, undue *influence of body weight or shape on self-evaluation, or denial of the seriousness of the current low body weight.*
D. Absence of at least three consecutive menstrual cycles *when otherwise expected to occur* (primary or secondary amenorrhea). (A woman is considered to have amenorrhea if her periods occur only following hormone, e.g., estrogen, administration.)	D. *In post-menarcheal females*, amenorrhea, i.e., the absence of at least three consecutive menstrual cycles. (A woman is considered to have amenorrhea if her periods occur only following hormone, e.g., estrogen, administration.)
	TYPE: *Restricting or Binge Eating/Purging*

(*continued*)

TABLE 1. (*continued*)

DSM-III-R	DSM-IV
Bulimia nervosa	
A. Recurrent episodes of binge eating (rapid consumption of a large amount of food in a discrete period of time).	A. Recurrent episodes of binge eating. An episode of binge eating is characterized by both of the following: (1) eating in a discrete period of time (e.g. within any two-hour period) *an amount of food that is definitely larger than most people would eat during a similar period of time and under similar circumstances*, and (2) a sense of lack of control over eating during the episode (e.g., a feeling that one cannot stop eating or control what or how much one is eating).
B. A feeling of lack of control over eating behavior during the eating binges.	B. *Recurrent inappropriate compensatory behavior* in order to prevent weight gain, such as: self-induced vomiting, misuse of laxatives, diuretics, *enemas or other medications, fasting or* excessive exercise.
C. The person *regularly engages in* either self-induced vomiting, use of laxatives or diuretics, strict dieting or fasting, or *vigorous* exercise in order to prevent weight gain.	C. The binge eating *and inappropriate compensatory behaviors* both occur, on average, at least twice a week for three months.
D. A minimum average of two binge eating episodes a week for at least three months.	D. *Self-evaluation is unduly influenced* by body shape and weight.
E. Persistent overconcern with body shape and weight.	E. *Disturbance does not occur exclusively during episodes of Anorexia Nervosa.*
	TYPE: *Purging or Nonpurging*

[a]Italics indicate differences between DSM-III-R and DSM-IV.

characterize the eating disorder patient and that are at the extreme end of an eating behavior continuum, such as obsessive eating patterns or a preoccupation with weight and body shape. However, the authors discussed data indicating that deficits in other areas, such as perceptual abilities and ego formation, represent behaviors that are uncorrelated with normal dieting. Polivy and Herman stressed that further research is necessary to resolve this issue.

The second controversy in case definition involves the determination of subtypes of anorexia and bulimia. Although the symptoms of bulimia and anorexia are distinct, they may co-occur within the same individual (Schlesier-Stropp, 1984). Because they may, it has been suggested that anorexia nervosa should include both restricting behaviors and binge eating/purging behaviors that lead to a reduction in body weight. Bulimia, in contrast, would involve binge eating/purging behavior not associated with weight change (DaCosta & Halmi, 1992; Walsh, 1992).

Johnson and Connors (1987) supported a different approach to subdividing the eating disorders. They contended that the distinction between bulimic and restricting behaviors is more useful clinically than the distinction between the diagnoses of bulimia and anorexia with binge eating/purging behavior. They suggested that rather than basing diagnoses on body weight, a differential diagnosis between anorexia and bulimia should depend on the presence or absence of binge eating.

These different approaches to diagnosing the eating disorders are reflected in the two most recent editions of the *Diagnostic and Statistical Manual of Mental Disorders* (DSM-III-R and DSM-IV) (APA, 1987, 1994), shown in Table 1. DSM-III-R describes specific behaviors that differentiate normal dieting from eating disorders, including maintenance of low body weight (15% below minimum expected weight), intense fear of gaining weight despite being underweight, exaggerated perception of body size and shape, and use of vigorous exercise, strict dieting, drugs, or vomiting to prevent weight gain. These abnormal eating behaviors are then divided into the two conventional categories, anorexia nervosa, requiring a reduction in body weight, and bulimia nervosa, requiring chronic binge eating and subsequent purging that maintains a normal to subnormal weight. This third edition—revised of the DSM requires a dual diagnosis for patients who engage in both types of behaviors.

In contrast to DSM-III-R, DSM-IV subdivides anorexia nervosa into either a "restricting" or a "binge eating/purging" type, while bulimia nervosa is divided into either a "purging" or a "nonpurging" type. Requirements for a diagnosis of bulimia were also modified to exclude people who exhibit bulimic behaviors exclusively during episodes of anorexia. Future etiological studies of the eating disorders that utilize DSM-IV criteria should help to clarify the usefulness of the recently defined subtypes.

EPIDEMIOLOGY

Estimates of prevalence rates of the eating disorders vary widely in the literature, with some studies noting rates as high as 5% for anorexia (Szmukler, 1985) and 19% for bulimia in females (Halmi, Falk, & Schwartz, 1981). Most of the more elevated estimates, however, resulted from studies in which the definition of the disorders was less strict than the DSM-III-R criteria. DSM-IV summarizes prevalence rates derived from DSM-III-R criteria ranging from 0.5% to 1.0% for anorexia and from 1% to 2% for bulimia in adolescent and young adult women (APA, 1994). The comparable rates for males are 10–20 times *less* than for females (Hoek, 1993; Hsu, 1990).

Symptoms of anorexia and bulimia typically emerge in early adulthood. Studies of clinic populations suggest a mean onset of 17 years for anorexia

(APA, 1994; Hsu, 1990; Shisslak, Crago, Neal, & Swain, 1987) and 18 years for bulimia (APA, 1987; Schlesier-Stropp, 1984). Although the age of onset for the eating disorders is rarely younger than 14 years (Lask & Bryant-Waugh, 1992), the eating disorders appear to be more severe when early-onset disorder does occur (McKenzie & Joyce, 1992; Swift, 1982; Walford & McCune, 1991). Bunnell, Shenker, Nussbaum, Jacobson, and Cooper (1990) suggested that current diagnostic criteria are too strict to identify eating disorders in very young children, who may experience significant emotional and developmental difficulties even though they do not meet the criteria for a diagnosable disorder.

A variety of cultural correlates of anorexia and bulimia have also been identified. Members of more affluent social classes (Hoek, 1993), those living in Western societies or rapidly changing cultures (Mumford, 1993), and individuals from special populations, such as ballet students, fashion models, and athletes (Hoek, 1993; Hsu, 1990; Szmukler, 1985), are all more likely than members of the general population to suffer from eating disorders. Caution must be exercised, however, in interpreting these data. For example, as Mumford (1993) pointed out, prevalence rates of eating disorders in non-Western cultures have been derived from questionnaires developed for use in Western populations. Because of potential cultural differences in language use and expression of symptoms, questionnaires must be validated in the specific population before the data can be interpreted. With similar concern, the apparent 2-fold increase in the incidence of eating disorders between 1965 and 1985 (Szmukler, 1985) must be further examined. Although some have suggested that a cultural emphasis on fitness and low body fat may result in more extreme dieting and disordered eating behavior, improvements in diagnostic criteria and heightened awareness of the disorder may account for much of the increase in the incidence of eating disorders.

DEVELOPMENTAL MODELS FOR THE EATING DISORDERS

Several models describing the impact of biological, psychological, and social factors on the development of anorexia and bulimia nervosa have been proposed over the past 15 years (Garfinkel & Garner, 1982; Johnson & Connors, 1987; Lucas, 1981; Tobias, 1988; Strober, 1991a). Each of these models emphasizes different relevant etiological factors, although they all suggest that these factors interact over time to produce disordered eating patterns. In Fig. 1, we have attempted to summarize the common themes from the developmental models and have indicated that genes and environment may impact vulnerability factors and stressors throughout development.

Vulnerability Factors

Several dimensions of vulnerability that might account for factors in the first box in Fig. 1 have been suggested for both anorexia and bulimia nervosa. With

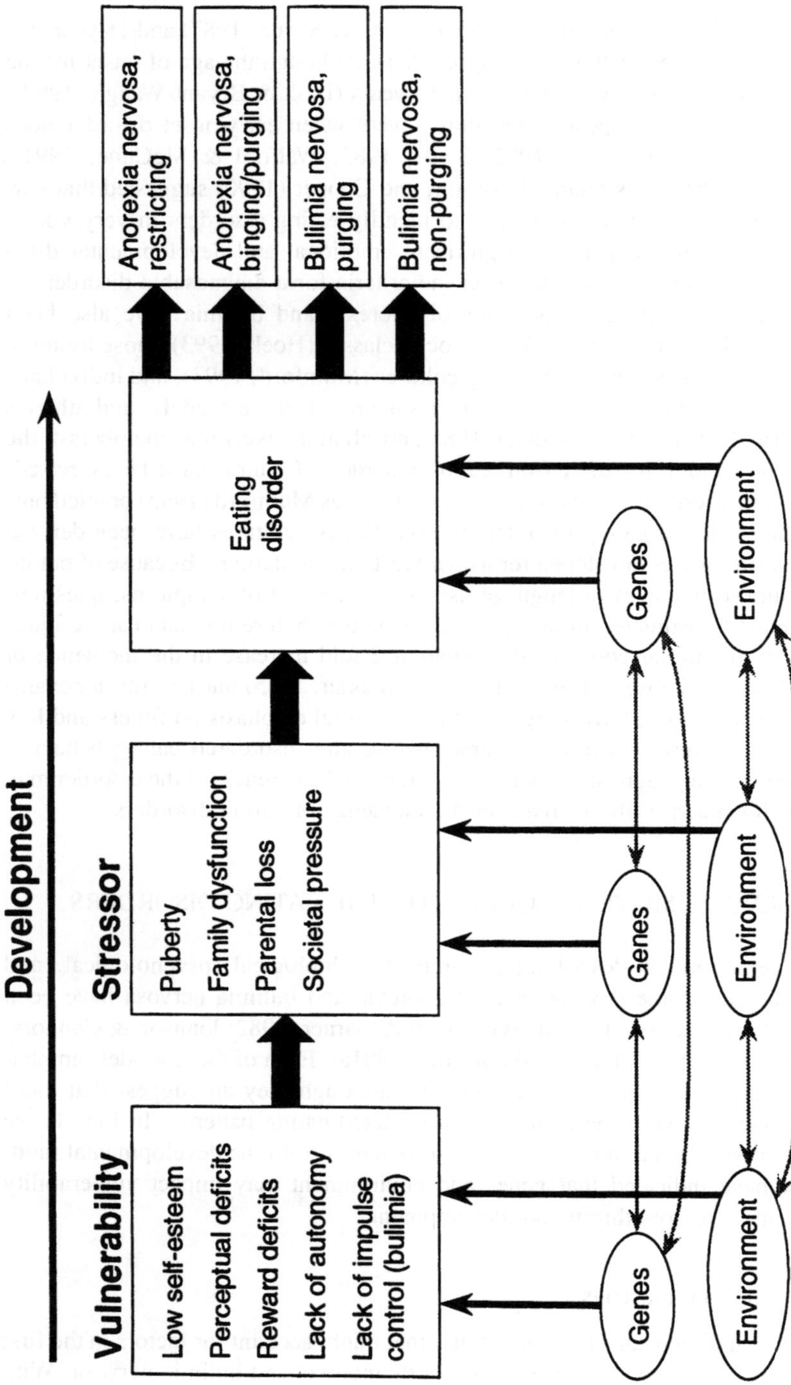

FIGURE 1. A developmental model for the eating disorders.

regard to anorexia, patients differ from controls in having a greater lack of autonomy and identity (Bruce, 1973; Selvini-Palazzoli, 1978), lower self-esteem (Grant & Fodor, 1986), perceptual deficits (Andersen & Hay, 1985; Gardner & Moncrieff, 1988), and an over-controlled ego (Kerr, Skok, & McLaughlin, 1991). It has been suggested that individuals with these characteristics, particularly females, lack the psychological readiness for the demands placed on them by puberty (Crisp, 1978). During this time, the increasing weight associated with puberty is linked to the societal demands of maturation. As a result of this linkage, anorexic patients feel they can control their rapidly maturing social image by asserting excessive control over their weight.

Bruch (1977) suggested that these patients experience an "all-or-none" thought process, in which the response to society's expectations for thinness is to severely restrict food intake. Although this idea is provocative, there are few data currently available to support the idea. Another theory has emerged (Strober, 1991a) suggesting that anorexia can be explained in part by the covariation of three personality domains outlined by Cloninger (1988). These domains include harm avoidance, novelty seeking, and reward dependence, which are all under partial genetic control. Strober suggested that when individuals are predisposed to avoid novelty and harm and to seek rewards, they may experience a need to maintain strict control over their own body size. This theory yields some testable hypotheses that may relate genetic influences on these personality domains to genetic influences on eating disorders.

Depressive symptoms also appear to be common among individuals with anorexia, especially those with early onset (Lask & Bryant-Waugh, 1992). It remains debatable, however, whether depressive symptoms are a cause, a consequence, or simply a correlate of the eating disorders (Herpertz-Dahlmann & Remschmidt, 1990). We return to this topic later in the chapter when we review several studies in the literature that have examined the familial association between eating disorders and depression. Finally, individual characteristics that are commonly associated with anorexia, but are less likely to represent causal factors, include obsessions with food and sexual inhibitions (Coovert, Kinder, & Thompson, 1989; Kerr et al., 1991; Swain, Shisslak, & Crago, 1991).

Issues of autonomy, identity, and low self-esteem, as discussed in regard to anorexia, have also been postulated as etiological factors in bulimia. Furthermore, bulimic patients have been shown to have difficulty with impulse control, identifying internal states, and controlling their emotions (Halmi, 1987; Johnson & Connors, 1987). Johnson and Connors suggested that a sense of helplessness arises when emotions cannot be controlled and leads the individual with bulimia to attempt to control at least part of his or her bodily experience. There has also been some indication that the susceptibility to binge eating may reflect unsuccessful attempts to maintain a body weight below the biological set point through continuous dieting. It is theorized that this dieting results in a state of deprivation that then triggers an excessive appetite to restore body fat to its set-point level (for a review, see Tobias, 1988).

If supported empirically, this explanation would constitute one of the few physiological mechanisms for the eating disorders.

As with anorexia, a linear relationship between the severity of bulimic behaviors and the severity of depressive symptoms has been indicated (Wilson & Lindholm, 1987). Individuals with bulimia also tend to have higher levels of anxiety, as compared to controls (Halmi, 1987), although the causal nature of this relationship is not known.

Stressors

Based on the Johnson and Connors (1987) conceptualization, the second box in Fig. 1 displays the influence of stressors that might precipitate the onset of an eating disorder in a vulnerable individual. Among the many hypothesized stressors are the onset of puberty, sociocultural factors such as an emphasis on dieting and thinness, and acute or chronic life events, including family dysfunction. A great deal of emphasis has been placed on the role of the family in the development of anorexia nervosa (Minuchin, Rosman, & Baker, 1978); the data suggest that families of anorexic patients exhibit behaviors, such as abnormally close relationships (enmeshment) and deviant communication patterns (conflict avoidance, parental overprotection), that are not present in families of normal controls (Crisp, Hsu, Harding, & Hartshorn, 1980; Kerr et al., 1991; Minuchin et al., 1975, 1978; Selvini-Palazzoli, 1978; Strober & Humphrey, 1987). An actual or perceived separation or loss of family members (through divorce, death, or parental infidelity) may also mediate the onset of anorexia (Kalucy, Crisp, & Harding, 1977; Theander, 1970).

Families of bulimic patients exhibit characteristics that are slightly different from those of families of individuals diagnosed with anorexia. In general, families of bulimic patients appear to have more hostility and less warmth than others (Hodes, 1993; Kog & Vandereycken, 1985). Marchi and Cohen (1990) provided longitudinal data indicating that conflictual behavior that occurred specifically around meals was also associated with bulimic symptoms later in life.

Synthesis

The common theme of developmental models for the eating disorders is that both vulnerabilities and stressors are required for onset. It is possible that unique combinations of predisposing and precipitating factors may define more homogeneous pathways to developing a disorder, although these pathways have not yet been elucidated. We also know relatively little about the genetic etiology of vulnerabilities and stressors. It is important to realize, however, that genes can influence different aspects of the developmental pathways. Genetic influences may impact individual factors, such as self-esteem or personality, and family stressors, such as overprotection. Furthermore, genetic factors may control an individual's sensitivity to environmental stressors that precipitate eating disorders, creating gene–environment interactions (cf., Kendler & Eaves, 1986). Ultimately, genetic influences on developmental precursors of anorexia or bulimia nervosa will result in familial similarity for the disorders themselves. The majority of our knowledge regarding genetic factors in the eating disorders has stemmed from investigation of this similarity. We now review the existing research in this area.

FAMILIAL TRANSMISSION OF THE EATING DISORDERS

Overview

Compared to findings in more prevalent psychiatric disorders such as major depression or alcoholism, there are relatively few data on the familial aggregation of eating disorders. What is known comes from two lines of evidence: family studies and twin studies. Each type of study has its advantages and drawbacks, and some discussion of these approaches is warranted here.

When studying rare diseases or psychiatric disorders, a family study is more often used than the twin design because the co-occurrence of two rare events (twinning and disorder) is not required to define a proband. Furthermore, in a family study, psychiatric data are typically obtained on *several* first- and second-degree relatives. This procedure provides more information, on a per-proband basis, than a twin study. However, as is often pointed out, the coaggregation of a disorder in a family may be due to shared genes or shared features of the family environment. The family study cannot disentangle these effects. The discrimination between genetic and family environmental influences is especially important in studying eating disorders, since family dysfunction has been implicated as a risk factor for development of the disorders (for reviews, see above and Strober & Humphrey, 1987; Strober, 1991b).

Unlike a family study, the twin method does allow one to determine whether familial aggregation is due to shared genes, shared family environment, or both. Unfortunately, this determination has not always been made in twin studies of anorexia and bulimia. Instead, the monozygotic (MZ) and dizygotic (DZ) concordance rates are simply compared, and if the former exceeds the latter, genetic influences are implicated. In only one case (Kendler et al., 1991) has a comparative model-fitting analysis been conducted to identify the factors that best account for familial aggregation.

Another positive attribute of the twin design is that it inherently controls for cohort effects on the prevalence of anorexia and bulimia (Bushnell, Wells, Hornblow, & Oakley-Brown, 1990; Kendler et al., 1991). These influences will serve to diminish familial correlations *across* generations.

One potential drawback of using twins to characterize the genetic contributions to eating disorders is that the twin relationship itself has been implicated as a risk factor for anorexia and bulimia, since it may hinder the development of a personal identity (Fichter & Noegel, 1990). There is no evidence, however, to suggest that the prevalence of eating disorders in the twin population is any higher than the prevalence in the non-twin population (Simonoff, 1992).

Family Studies

A concise summary and comparison of family studies of anorexia and bulimia presents difficulties, since the investigations differ in several ways, including proband ascertainment, diagnostic criteria, method of assessing illness in the relatives, and characteristics of the control population. In general, the evidence from five family studies of moderate size indicates that eating disorders are more frequent in the relatives of probands. However, since the general population risk for eating disorders is relatively low, some studies have not had the power to detect 2- or 3-fold increases in risk among the patients of relatives. Below, we summarize the major findings from these five investigations.

Gershon and colleagues (Gershon, Schreiber, & Hamovit, 1983; Gershon et al., 1984) conducted the first family study of anorexia nervosa in 99 relatives of 24 patients and 265 relatives of 43 controls. When possible, diagnoses on the majority of the relatives were made through direct interview [Schedule for Affective Disorders and Schizophrenia—Lifetime Version (SADS-L)] (Mazure & Gershon, 1979), rather than through the report of an informant. The former method of data collection has been shown to have greater sensitivity for case detection. Gershon and colleagues considered anorexia and bulimia separately and found no cases of anorexia in the relatives of controls, while the lifetime risk of anorexia in the relatives of the patients was 2%. The comparative figures for bulimia were 1.3% in the controls and 4.4% in the relatives of the probands.

These data argue for a nonspecific, family influence on the development of anorexia and bulimia. Lifetime risks similar to these were obtained by Strober, Lampert, Morrell, & Burroughs (1990) in a larger study of 97 anorexic patients and their 387 first-degree relatives. When compared to the 700 relatives of the controls, in whom the prevalence of eating disorders was 1.1%, relatives of probands were at an increased risk for both anorexia (4.1%) and bulimia (2.6%).

In 1987, Hudson, Pope, Jonas, Yurgelun-Todd, and Frankenburg (1987) presented data from a family history study of 69 bulimic patients, their 283 first-degree relatives, and the 149 relatives of 28 nonpsychiatric controls. Similar to the Gershon et al. (1984) study, these investigators considered the prevalence of both anorexia and bulimia in the first-degree relatives. For both diagnoses, lifetime risk for the relatives of probands was increased over that of the control relatives (2.2% vs. 0%). However, these differences did not reach statistical significance.

An additional family study of bulimia nervosa was reported in 1989 by Kassett et al. (1989). These investigators used the SADS-L to obtain lifetime diagnoses of the eating disorders in 185 relatives of 40 probands and 118 relatives of matched controls. Results indicated a marked elevation of anorexia and bulimia in the relatives of patients with bulimia (lifetime risks 2.2% and 9.6%, respectively) compared to control families (0% and 3.5%). This difference in lifetime risks was significant when the eating disorders were combined in a single diagnostic category.

One investigation that failed to replicate the findings of the family studies was described by Logue, Crowe, and Bean (1989). Their analysis of eating disorders in 146 relatives of 30 patients (diagnosed with both anorexia and bulimia) found no cases of either disorder, nor were any diagnoses made in 107 relatives of normal controls. Simply by chance, however, one could fail to detect a family risk of 2–4% in this relatively small sample size of relatives.

In summary, the data from family studies suggest that eating disorders aggregate in the families of patients with anorexia and bulimia nervosa. For females, there is a 2- to 3-fold increase in the lifetime risk of developing an eating disorder if a first-degree relative is affected. However, the *absolute* risk that such an individual will develop an eating disorder is fairly small, given the low prevalence of anorexia and bulimia nervosa in the population.

Twin Studies

As with family studies, data on the coaggregation of the eating disorders in twins are limited. Although several case reports on twin pairs concordant and discordant for the eating disorders have appeared in the literature (for a review, see Garfinkel & Garner, 1982), the pooling of these data is questionable due to various methods of twin ascertainment, psychiatric diagnosis, and zygosity diagnosis. Recently, efforts have been made to systematically ascertain affected twins from clinic populations, or to utilize population-based twin registries, to estimate twin concordance rates. Four examples are provided in Table 2, along with corresponding concordance rates for different eating disorder diagnoses.

In the Treasure and Holland (1990) report, twins were included in the study if they had received treatment for their eating disorder and were ascertained through eating disorder units within the United Kingdom or through the Maudsley Twin Register. The authors reported twin concordance rates for four different diagnostic classifications: any eating disorder, restricting anorexia, bulimia with or without anorexia, and all cases of anorexia. For each classification, the MZ concordance rate exceeded the DZ concordance rate, indicating that genetic factors play a role in the development of eating disorders. Moreover, the genetic factors appeared to be more important for anorexia than for bulimia. However, the authors did not statistically evaluate the significance of this difference.

The study by Fichter and Noegel (1990) ascertained twins with bulimia nervosa from clinic populations and through a press survey about bulimia. Their analysis of 21 pairs of female twins indicated that MZ twins were highly concordant for both bulimia and the eating disorders combined. MZ twin concordance exceeded that of DZ twins more than 2-fold, a significant difference. Genetic influences on bulimia nervosa were also implicated in a very small twin study (11

TABLE 2. Female Twin Concordance Rates for Eating Disorders

Study	n Pairs	Ascertainment	Diagnosis	Concordance MZ	Concordance DZ
Treasure & Holland (1990); Holland, Hall, Murray, Russell, & Crisp (1984)	31 MZ, 28 DZ	Clinic	All eating disorders	55%	27%
			Restricting anorexia	66%	0%
			Bulimia with & without anorexia	35%	29%
			Anorexia	68%	8%
Fichter & Noegel (1990)	6 MZ, 15 DZ	Clinic, press survey	Bulimia	83%	27%
			All eating disorders	83%	33%
Hsu et al. (1990)	6 MZ, 5 DZ	Clinic	Bulimia	33%	0%
Kendler et al. (1991)	590 MZ, 440 DZ	Population-based twin registry	Narrowly defined bulimia	23%	8.7%
			Broadly defined bulimia	26%	16%

pairs) by Hsu, Chesler, and Santhouse (1990), in which two of the MZ pairs, but none of the DZ pairs, were concordant.

The largest reported twin study of bulimia nervosa is that by Kendler et al. (1991), utilizing the population-based Virginia Twin Registry. Of the 2163 individual twins interviewed, 60 were diagnosed as having *definite* or *probable* bulimia nervosa (met DSM-III-R criteria) and 63 were diagnosed with *possible* bulimia. The twin concordance rates for these two diagnostic classifications (Table 2) indicate a significantly higher concordance rate among MZ twins. However, the MZ concordance is reduced compared to other (clinic) samples of twins. This difference may reflect an ascertainment bias in which treatment-seeking twins with bulimia are more likely to be concordant for the disorder.

In their analysis of the twin data, Kendler and colleagues also fit genetic models to the twin correlations for liability to disorder. They concluded that the best-fitting model to the twin data contained only additive genetic and individual environmental effects, with genetic effects explaining 52% and 55% of the variance for the broad and narrow definitions of bulimia, respectively. Thus, in the largest twin study conducted to date, there was no evidence that the shared family environment influenced twin similarity. However, it must be kept in mind that this finding does not necessarily imply that family features, including dysfunction and discord, are not important in the development of the eating disorders. Instead, the result indicates that individuals within a family *are not affected in the same way* by characteristics of their shared environment. The possibility remains that individuals within a family respond differently to the same environmental conditions.

Familial Nature of Comorbidity

Earlier in this chapter, we discussed the co-occurrence of eating disorders with symptoms of depression. There are several hypotheses that could explain this comorbidity, including (1) common causal factors, which may be genetic or environmental in origin; (2) one disorder preceding, and serving as a direct risk factor for, the development of a second disorder [e.g., Major Depressive Disorder (MDD) leading to an eating disorder]; and (3) the physiological consequences of one disorder resulting in a second disorder (e.g., starvation resulting in MDD). Under each of these hypotheses, different sets of predictions can be made about the co-occurrence of disorders within individuals and within families. For example, if hypothesis (1) holds, and there is a genetic correlation between the eating disorders and MDD, then we would expect the rates of both MDD and the eating disorders to be elevated in the relatives of both types of probands. The extent of this elevation will depend on the genetic correlation between the two disorders and the population prevalence rates. In contrast, under a causal hypothesis, presence of the causative disorder in the proband predicts an increase of both disorders in the relatives, while presence of only the disorder of consequence does not. Of course, it may be that neither a causal hypothesis nor a common factor hypothesis explains all cases of comorbidity. Each mechanism may account for some comorbid cases, and each may help to define more homogeneous groups of eating disorder patients.

In practice, it becomes difficult to investigate the nature of comorbidity because of the sample sizes, and the genetic design, that are required to discriminate between competing hypotheses. However, some information on the familial association of affective disorders and eating disorders is available from family studies discussed earlier. Investigations by Gershon et al. (1984), Hudson et al. (1987), Kassett et al. (1989), and Logue et al. (1989) have examined the risk of Major Affective Disorders (MADs) in the first-degree relatives of probands with bulimia or anorexia. Of concern in these investigations was (1) whether rates of affective illness were elevated in the relatives of the probands and (2) whether this elevation was unique to those probands who had a history of a MAD. The latter suggestion, if supported by the data, would indicate that comorbid probands had a unique family history profile. Consequently, it could be hypothesized that the comorbid probands and individuals with only MDD may respond more similarly to treatment than those probands who present with only an eating disorder.

A summary of the family studies of comorbidity is presented in Table 3. We show the risk of MDD in the relatives of the probands and controls. In some of

TABLE 3. Risk of Major Depression in the Relatives of Eating Disorder Probands[a]

Study	Proband	Number of relatives	Risk of MDD in relatives[b]	Comments
Logue et al. (1989)	Eating disorder with MDD	46	13	No difference between eating disorder with and without MDD.
	Eating disorder without MDD	86	12	Difference between eating disorder and controls.
	MDD	100	7	
	Control	75	5	
Kassett et al. (1989)	Bulimia with MAD	100	29.5	Bulimia with MAD different from controls.
	Bulimia without MAD	85	12.5	No difference between bulimia without MAD and controls.
	Control	118	8.8	
Hudson et al. (1987)	Bulimia with MAD	283	36.2	Difference between bulimia with and without MAD.
	Bulimia without MAD		18.7	
	MAD	104	24.6	
	Control	149	6.7	No difference between bulimia without MAD and MAD. Difference between controls and all other groups.
Gershon et al. (1984)	Anorexia with MAD	53	13.4	No difference between anorexia or bulimia with and without MAD.
	Anorexia without MAD	46	13.1	
	Bulimia with MAD	48	15.8	
	Bulimia without MAD	51	11.1	
	Control	265	5.8	

[a]Abbreviations: (MDD) major depressive disorder; (MAD) major affective disorder.
[b]Unadjusted risk is presented by Logue et al. (1989); lifetime risk is cited in the remaining studies.

the investigations, subjects with a MAD are included in the analysis as an additional control.

In each of these family studies, the risk of MDD in the relatives of eating disorder patients exceeds that for control relatives. However, in only two cases did eating disorder patients with a lifetime diagnosis of a MAD have a greater family history of MDD than those individuals with an eating disorder only. These findings suggest a common familial risk for the development of MDD and eating disorders, but the evidence is equivocal as to whether comorbidity in the eating disorder patient serves as an index of greater familial vulnerability to the development of affective disorders.

Walters, et al. (1992) have used a twin design to determine whether the familial co-occurrence of eating disorders and affective disorders is due to shared genes or shared environment. Their report indicated that the co-occurrence of bulimia and MDD in 1033 pairs of female twins from the general population is mediated entirely by genes. Environmental influences, unique to the individual or shared between twins, do not contribute to comorbidity. This result would imply that the increase in affective disorders in relatives of eating disorder patients is due to a higher genetic loading (influencing both disorders) in the family. Certainly, additional family and twin studies are needed to replicate these findings and to address issues of causation.

GENETIC STUDIES OF PSYCHOLOGICAL DOMAINS RELATED TO EATING DISORDERS

Although the vast majority of twin studies on the eating disorders have focused on diagnostic entities, there is one investigation deserving mention that has considered genetic influences on continuous measures of psychological domains related to bulimia and anorexia. Rutherford, McGuffin, Katz, and Murray (1993) analyzed data from the Eating Attitudes Test (EAT) (Garner & Garfinkel, 1979) and the Eating Disorder Inventory (EDI) (Garner, Olmsted, and Garfinkel, 1983) obtained on 147 MZ and 99 DZ female twin pairs in a volunteer-based twin registry. Both instruments have been used extensively in epidemiological surveys of behaviors and personality traits that characterize eating disorders. In their study, Rutherford et al. (1993) used a shortened version (26 items) of the EAT, which includes items regarding preoccupation with food and dieting. A factor analysis of the EAT yielded a single factor, which explained 31% of the total variance. The EDI, in contrast, is a lengthier instrument (64 items) with eight subscales: Drive for Thinness, Bulimia, Body Dissatisfaction, Ineffectiveness, Perfectionism, Interpersonal Distrust, Interoceptive Awareness, and Maturity Fears. In validation studies (Garner et al., 1983), patients with anorexia differed from normal controls on all subscales. Further, patients with co-occurring symptoms of bulimia and anorexia scored higher on the Bulimia and Body Dissatisfaction subscales than the patients with only anorexia, while a comparison group of bulimic patients scored lower than patients with either restricting *or* binge eating/purging anorexia on the Maturity Fears scale.

Genetic analyses were conducted on the single EAT factor and the 8 subscales of the EDI to determine the extent to which normal variation in eating behavior is influenced by genetic effects. The investigators adopted a maximum-likelihood model-fitting approach for the data analysis, so that they could evaluate the significance of genetic, shared environmental, and individual environmental influences on the scale scores. Results indicated a moderate heritability of all subscales [ranging from 25% (for the EDI Interpersonal Distrust subscale) to 52% (for Body Dissatisfaction)], with no evidence for a significant effect of the family environment. Moreover, for all but three subscales (Bulimia, Perfectionism, and Maturity Fears), the DZ twin correlation was less than half the MZ twin correlation, suggesting that nonadditive genetic effects play a role in these personality traits and behaviors characteristic of eating disorder patients.

Although not considered by Rutherford and colleagues, an informative extension of this study would be a multivariate genetic analysis of the subscales along with eating disorder diagnoses (cf., Neale & Cardon, 1992). This method would allow the investigators to determine whether the same or different genetic factors are operating on the continuous measures of eating disorder liability. Additionally, the extent of genetic variation in the diagnoses themselves, which are explained by the subscales, could be evaluated.

FUTURE DIRECTIONS

The paucity of data on the genetics of the eating disorders leaves room for several outstanding issues to be addressed and current findings to be replicated. Among the questions that future studies could investigate are:

1. How are anorexia nervosa and bulimia nervosa related genetically?
2. Is the current (DSM-IV) nosological separation of binge eating/purging anorexia vs. bulimia supported by family–genetic studies?
3. How do genes influence vulnerabilities and stressors associated with the eating disorders?
4. Is there evidence for a gene–environment interaction for the development of eating disorders?
5. Do inherited vulnerabilities to eating disorders express themselves differently in females and males?

Determining the answers to these questions will require genetically informative samples with the power to discriminate among competing hypotheses. Furthermore, longitudinal genetic studies, utilizing indices of liability to disorder, will be of particular use in characterizing developmental pathways that lead to an eating disorder. It is highly unlikely that any single genetic mechanism will explain the familial nature of anorexia and bulimia nervosa. Instead, future studies should help to define a number of ways in which inherited characteristics can predipose individuals to an eating disorder.

REFERENCES

American Psychiatric Association (1980). *Diagnostic and statistical manual of mental disorders*, 3rd ed. Washington, DC: American Psychiatric Press.

American Psychiatric Association (1987). *Diagnostic and statistical manual of mental disorders*, 3rd ed.—rev. Washington, DC: American Psychiatric Press.

American Psychiatric Association (1994). *Diagnostic and statistical manual of mental disorders*, 4th ed. Washington, DC: American Psychiatric Press.

Andersen, A. E., & Hay, A. B. (1985). Racial and socioeconomic influences in anorexia nervosa and bulimia. *International Journal of Eating Disorders, 4*, 479–487.

Bruce, H. (1973). *Eating disorders: Obesity, anorexia nervosa and the person within*. New York: Basic Books.

Bruch, H. (1977). Anorexia nervosa. In E. D. Wittkower and H. Warnes (Eds.), *Psychosomatic medicine: Its clinical applications* (pp. 229–237). New York: Harper & Row.

Bunnell, D. W., Shenker, I. R., Nussbaum, M. P., Jacobson, M. S., & Cooper, P. (1990). Subclinical versus formal eating disorders: Differentiating psychological features. *International Journal of Eating Disorders, 9*, 357–362.

Bushnell, J. A., Wells, J. E., Hornblow, A. R., & Oakley-Brown, M. A. (1990). Prevalence of three bulimia syndromes in the general population. *Psychological Medicine, 20*, 671–680.

Cloninger, C. R. (1988). A unified theory of personality and its role in the development of anxiety states: A reply to commentaries. *Psychiatric Developments, 6*, 83–120.

Coovert, D. L., Kinder, B. N., & Thompson, J. K. (1989). The psychosexual aspects of anorexia nervosa and bulimia nervosa: A review of the literature. *Clinical Psychology Review, 9*, 169–180.

Crisp, A. H. (1978). Some aspects of the relationship between body weight and sexual behavior with particular reference to massive obesity and anorexia nervosa. *International Journal of Obesity, 2*, 17–32.

Crisp, A. H., Hsu, L. K. G., Harding, B., & Hartshorn, J. (1980). Clinical features of anorexia nervosa: A study of a consecutive series of 102 female patients. *Journal of Psychosomatic Research, 24*, 179–191.

DaCosta, M., & Halmi, K. A. (1992). Classifications of anorexia nervosa: Question of subtypes. *International Journal of Eating Disorders, 11*, 305–313.

Fichter, M. M., & Noegel, R. (1990). Concordance for bulimia nervosa in twins. *International Journal of Eating Disorders, 9*, 255–263.

Gardner, R. M., & Moncrieff, C. (1988). Body image distortion in anorexics as a non-sensory phenomenon: A signal detection approach. *Journal of Clinical Psychology, 44*, 101–107.

Garfinkel, P. E., & Garner, D. M. (1982). *Anorexia nervosa: A multidimensional perspective*. New York: Brunner/Mazel.

Garner, D. M., & Garfinkel, P. E. (1979). The Eating Attitudes Test: An index of the symptoms of anorexia nervosa. *Psychological Medicine, 9*, 273–279.

Garner, D. M., Olmsted, M. P., & Garfinkel, P. E. (1983). Does anorexia nervosa exist on a continuum? *International Journal of Eating Disorders, 2*, 11–20.

Gershon, E. S., Schreiber, J. L., & Hamovit, J. R. (1983). Anorexia nervosa and major affective disorders associated in families: A preliminary report. In S. B. Guze, F. Earls, & J. E. Barrett (Eds.), *Childhood psychopathology and development* (pp. 279–286). New York: Raven Press.

Gershon, E. S., Schreiber, J. L., Hamovit, J. R., Dibble, E. D., Kaye, W., Nurnberger, J. I., Andersen, A. E., & Ebert, M. (1984). Clinical findings in patients with anorexia nervosa and affective illness in their relatives. *American Journal of Psychiatry, 141*, 1419–1422.

Grant, C. L., & Fodor, I. G. (1986). Adolescent attitudes toward body image and anorexic behavior. *Adolescence, 21*, 269–281.

Halmi, K. A. (1987). Anorexia nervosa and bulimia. In V. B. VanHasselt & M. Hersen (Eds.), *Handbook of adolescent psychology* (pp. 265–287). New York: Pergamon Press.

Halmi, K. A., Falk, J. R., & Schwartz, E. (1981). Binge-eating and vomiting: A survey of a college population. *Psychological Medicine, 11*, 697–706.

Herpertz-Dahlmann, B., & Remschmidt, H. (1990). Anorexia nervosa and depression: A continuing debate. In H. Remschmidt & M. H. Schmidt (Eds.), *Child and youth psychiatry: European perspectives* (pp. 69–84). Stuttgart: Hogrefe & Huber.

Hodes, M. (1993). Anorexia nervosa and bulimia nervosa in children. *International Review of Psychiatry, 5*, 101–108.

Hoek, H. W. (1993). Review of the epidemiological studies of eating disorders. *International Review of Psychiatry, 5*, 61–74.

Holland, A. J., Hall, A., Murray, R., Russell, G. F. M., & Crisp, A. H. (1984). Anorexia nervosa: A study of 34 twin pairs and one set of triplets. *British Journal of Psychiatry, 145*, 414–419.

Hsu, L. K. G. (1989). The gender gap in eating disorders: Why are the eating disorders more common among women? *Clinical Psychology Review, 9*, 393–407.

Hsu, L. K. G. (1990). *Eating disorders*. New York: Guilford Press.

Hsu, L. K. G., Chesler, B. E., & Santhouse, R. (1990). Bulimia nervosa in eleven sets of twins: A clinical report. *International Journal of Eating Disorders, 9*, 275–282.

Hudson, J. I., Pope, H. G., Jonas, J. M., Yurgelun-Todd, D., & Frankenburg, F. R. (1987). A controlled family history study of bulimia. *Psychological Medicine, 17*, 883–890.

Johnson, C., & Connors, M. E. (1987). *The etiology and treatment of bulimia nervosa: A biopsychosocial perspective*. New York: Basic Books.

Kalucy, R. C., Crisp, A. H., & Harding, B. (1977). A study of 56 families with anorexia nervosa. *British Journal of Medical Psychology, 50*, 381–395.

Kassett, J. A., Gershon, E. S., Maxwell, M. E., Guroff, J. J., Kazuba, D. M., Smith, A. L., Brandt, H. A., & Jimerson, D. C. (1989). Psychiatric disorders in the first-degree relatives of probands with bulimia nervosa. *American Journal of Psychiatry, 146*, 1468–1471.

Kendler, K. S., & Eaves, L. J. (1986). Models for the joint effect of genotype and environment on liability to psychiatric illness. *American Journal of Psychiatry, 143*, 279–289.

Kendler, K. S., MacLean, C., Neale, M., Kessler, R., Heath, A., & Eaves, L. (1991). The genetic epidemiology of bulimia nervosa. *American Journal of Psychiatry, 148(12)*, 1627–1637.

Kerr, J. K., Skok, R. L., & McLaughlin, T. F. (1991). Characteristics common to females who exhibit anorexic or bulimic behavior: A review of current literature. *Journal of Clinical Psychology, 47*, 846–853.

Kog, E., & Vandereycken, W. (1985). Family characteristics of anorexia nervosa and bulimia: A review of the research literature. *Clinical Psychology Review, 5*, 159–180.

Lacey, J. H. (1982). Anorexia nervosa and a bearded female saint. *British Medical Journal, 285*, 1816–1817.

Lask, B., & Bryant-Waugh, R. (1992). Early-onset anorexia nervosa and related eating disorders. *Journal of Child Psychology and Psychiatry and Allied Disciplines, 33(1)*, 281–300.

Logue, C. M., Crowe, R. R., & Bean, J. A. (1989). A family study of anorexia nervosa and bulimia. *Comprehensive Psychiatry, 30*, 179–188.

Lucas, A. R. (1981). Toward the understanding of anorexia nervosa as a disease entity. *Mayo Clinic Proceedings, 56*, 254–264.

Marchi, M., & Cohen, P. (1990). Early childhood eating behaviors and adolescent eating disorders. *Journal of the American Academy of Child and Adolescent Psychiatry, 29*, 112–117.

Mazure, C., & Gershon, E. S. (1979). Blindness and reliability in lifetime psychiatric diagnosis. *Archives of General Psychiatry, 36*, 521–525.

McKenzie, J. M., & Joyce, P. R. (1992). Hospitalization for anorexia nervosa. *International Journal of Eating Disorders, 11*, 235–241.

Minuchin, S., Baker, L., Rosman, B. L., Liebman, R., Milman, L., & Todd, T. C. (1975). A conceptual model of psychosomatic illness in children. *Archives of General Psychiatry, 32*, 1031–1038.

Minuchin, S., Rosman, B. L., & Baker, L. (1978). *Psychosomatic families: Anorexia nervosa in context*. Cambridge, MA: Harvard University Press.

Mumford, D. B. (1993). Eating disorders in different cultures. *International Review of Psychiatry, 5*, 109–113.

Neale, M. C., & Cardon, L. R. (1992). *Methodology for genetic studies of twins and families*. Dordrecht, The Netherlands: Kluwer Academic Press.

Nylander, I. (1971). The feeling of being fat and dieting in a school population: An epidemiologic interview investigation. *Acta Sociomedica Scandinavica, 1*, 17–26.

Polivy, J., & Herman, C. P. (1987). Diagnosis and treatment of normal eating. *Journal of Consulting and Clinical Psychology, 55,* 635–644.

Rutherford, J., McGuffin, P., Katz, R. J., & Murray, R. M. (1993). Genetic influences on eating attitudes in a normal female twin population. *Psychological Medicine, 23,* 425–436.

Schlesier-Stropp, B. (1984). Bulimia: A review of the literature. *Psychological Bulletin, 95(2),* 247–257.

Selvini-Palazzoli, M. (1978). *Self-starvation: From individual to family therapy in the treatment of anorexia nervosa.* New York: Jason Aronson.

Shisslak, C. M., Crago, M., Neal, M. E., & Swain, B. (1987). Primary prevention of eating disorders. *Journal of Consulting and Clinical Psychology, 55,* 660–667.

Simonoff, E. (1992). A comparison of twins and singletons with child psychiatric disorders: An item sheet study. *Journal of Child Psychology and Psychiatry and Allied Disciplines, 33,* 1819–1832.

Strober, M. (1991a). Disorders of the self in anorexia nervosa: An organismic–developmental paradigm. In C. L. Johnson (Ed.), *Psychodynamic treatment of anorexia nervosa and bulimia* (pp. 354–373). New York: Guilford Press.

Strober, M. (1991b). Family–genetic studies of eating disorders. *Journal of Clinical Psychiatry, 52,* 9–12.

Strober, M., & Humphrey, L. L. (1987). Familial contributions to the etiology and course of anorexia nervosa and bulimia. *Journal of Consulting and Clinical Psychology, 55(5),* 654–659.

Strober, M., Lampert, C., Morrell, W., & Burroughs, J. (1990). A controlled family study of anorexia nervosa: Evidence of familial aggregation and lack of shared transmission with affective disorders. *International Journal of Eating Disorders, 9,* 239–253.

Swain, B., Shisslak, C. M., & Crago, M. (1991). Clinical features of eating disorders and individual psychological functioning. *Journal of Clinical Psychology, 47,* 702–708.

Swift, W. J. (1982). The long-term outcome of early onset anorexia nervosa: A critical review. *Journal of the American Academy of Child Psychiatry, 21,* 38–46.

Szmukler, G. I. (1985). The epidemiology of anorexia nervosa and bulimia. *Journal of Psychiatric Research, 19,* 143–153.

Theander, S. (1970). Anorexia nervosa: A psychiatric investigation of 94 female cases. *Acta Psychiatric Scandinavia Supplementum, 214,* 1–194.

Tobias, A. L. (1988). Bulimia: An overview. in K. Clark, R. Parr, & W. Castelli (Eds.), *Evaluation and management of eating disorders: Anorexia, bulimia, and obesity* (pp. 173–186). Champaign, IL: Life Enhancement Publications.

Treasure, J., & Holland, A. (1990). Genes and the aetiology of eating disorders. In P. McGuffin & R. Murray (Eds.), *The new genetics of mental illness* (pp. 198–211). Oxford: Butterworth-Heinemann.

Walford, G., & McCune, N. (1991). Long-term outcome in early-onset anorexia nervosa. *British Journal of Psychiatry, 159,* 383–389.

Walsh, B. T. (1992). Diagnostic criteria for eating disorders in DSM-IV: Work is progress. *International Journal of Eating Disorders, 11,* 301–304.

Walters, E. E., Neale, M. C., Eaves, L. J., Heath, A. C., Kessler, R. C., & Kendler, K. S. (1992). Bulimia nervosa and major depression: A study of common genetic and environmental factors. *Psychological Medicine, 22,* 617–622.

Wilson, G. T., & Lindholm, L. (1987). Bulimia nervosa and depression. *International Journal of Eating Disorders, 6,* 725–732.

PART FIVE

GENDER AND ETHNICITY

CHAPTER TEN

Issues in the Behavior Genetic Investigation of Gender Differences

CHANDRA A. REYNOLDS AND JOHN K. HEWITT

INTRODUCTION

For many phenotypes of interest in behavioral medicine, there are differences between men and women. Preceding chapters discuss (1) alcoholism, the prevalence of which is markedly higher in men than in women; (2) smoking, which has historically occurred more often and at younger ages in men, though in recent years the prevalence has increased more rapidly among women; (3) cardiovascular indices, for which women tend to have higher resting heart rates and lower blood pressures; (4) body size, regarding which women are on average both shorter and lighter than men; and (5) eating disorders, which show dramatically higher prevalence in women than in men. Gender differences in these areas have generally been described in behavioral research in terms of average phenotypic differences (e.g., different prevalence rates). For some conditions, such as hemophilia or color blindness, higher prevalence rates in men than in women are a consequence of the fact that the gene that causes the condition is located on the X chromosome and there is a particular (recessive) form of gene action. However, most of the traits in which we are interested are not X-linked in this way, but are influenced by genes on the 22 pairs of (non-sex-determining) autosomes. In these cases, average gender differences are not immediately informative about gene action. But such average differences should at least alert us to the possibility that individual differences *among* men and *among* women might result from autoso-

---

CHANDRA A. REYNOLDS AND JOHN K. HEWITT • Institute for Behavioral Genetics, University of Colorado at Boulder, Boulder, Colorado 80309.

Behavior Genetic Approaches in Behavioral Medicine, edited by J. Rick Turner, Lon R. Cardon, and John K. Hewitt. Plenum Press, New York, 1995.

mal genes that have greater impact in women than in men (or vice versa), or that some genes that might contribute to vulnerability in women are distinct from those genes that might contribute to vulnerability in men.

These two possibilities are aspects of sex limitation of gene expression. Although sex limitation of gene expression has to date received little attention in behavioral medicine, a full understanding of risk factors in behavioral medicine must consider the possibility that genetic influences may play different roles in men and women. Of course, what may be true of genetic influences might equally be true for environmental risks, vulnerability, and protective factors; i.e., there may be sex limitation of environmental influences.

Genetically informative studies can provide the information necessary to distinguish differences in the magnitude and kind of genetic and environmental effects that contribute to individual variation. In this chapter, we first consider the importance of prevalence rates to the proper interpretation of data on family resemblance, often presented in the form of concordance rates for a disorder among pairs of relatives. Next, we see how these considerations extend to mean differences in quantitative traits. Finally, we return to a presentation of the available behavior genetic methods for the study of sex limitation in twin, family, and adoption studies. It is here that we feel that the most immediate progress can be made in the examination of gender differences in behavioral medicine research.

IMPORTANCE OF GENDER DIFFERENCES IN PREVALENCE RATES

Familial similarity for liability to a disorder or condition is usually reported in terms of concordance rates. These rates describe the probability that if one individual is affected, a relative of a given order will also be affected. However, these concordance rates are a function of *both* the underlying degree of familial resemblance *and* the prevalence rates of the disorder. Therefore, when interpreting genetically informative data, it is important to consider the prevalence rates for the population being studied. Three possible ways in which gender differences in concordance may be related to different prevalence rates between genders were reviewed by Cloninger, Christiansen, Reich, and Gottesman (1978).

In the first case, higher prevalence will result in higher observed concordance rates *for the same degree of underlying resemblance*. Thus, this model was termed the "isocorrelational" model (Cloninger et al., 1978). If we consider the concordance rates for alcoholism in pairs of twins, we expect the observed concordance for male pairs to be higher than that for female pairs, simply because the prevalence of alcoholism is higher in men than in women (see Chapter 2) (Heath, Slutske, & Madden, in press). To see why this should be so, consider Figs. 1–3, which depict possible liability distributions with men showing a greater prevalence than women for a given trait, such as alcoholism. By way of illustration, we are assuming an underlying normally distributed liability

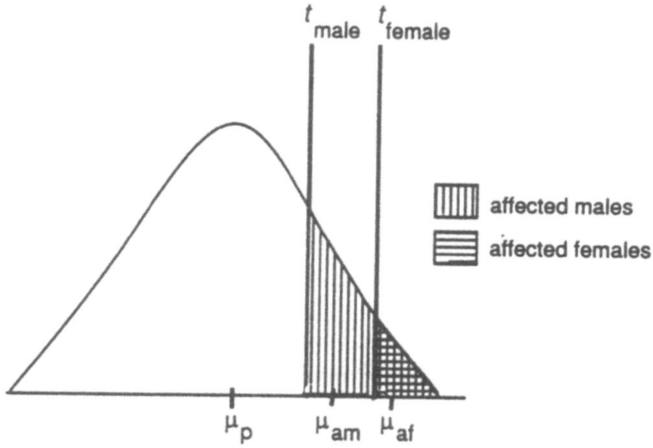

FIGURE 1. Multifactorial liability distribution with differing prevalence rates for males and females for alcoholism: isocorrelational model. Adapted from Cloninger et al. (1978).

for alcoholism; if an individual has a liability above the gender-specific threshold ($t_{males} < t_{females}$), then he or she becomes an alcoholic case. Figures 2 and 3 depict correlated liability distributions for dizygotic (DZ) twins. In these figures, quadrants b and c correspond to those pairs who are concordant for alcoholism and concordant for not being alcoholic, respectively, and quadrants a and d represent discordant pairs.

These three joint distributions are intended to depict identical correlations among the three twin types (same-sex male and female pairs and opposite-sex pairs). In clinically ascertained samples, we would identify an alcoholic individual, called the "proband," and then determine whether the proband's co-twin was also alcoholic. The concordance rate would correspond to the proportion b/(b + d) in Figure 2 and, in Figure 3, b/(b + d) or b/(a + b), depending on whether we have male or female probands. Following this procedure, DZ male twins show greater concordance than female DZ twins (Fig. 2), and opposite-sex DZ twins (Fig. 3) show a concordance that is contingent on the sex of the proband.

Clearly, these relationships do not necessarily imply greater genetic variance in one sex than in another. The same correlations underlie all three joint distributions, with differing prevalence rates for the two genders. In the isocorrelational model, the females show greater deviation in liability and their relatives are more deviant as a group than the relatives of affected male probands. Thus, there are proportionately more affected relatives for female probands than for male probands (see Fig. 3), but there is the same magnitude of environmental and genetic factors operating in both sexes. This model cannot explain the differences in prevalence rates that occur phenotypically, though it can be used to test whether or not same-sex and opposite-sex relatives are equally correlated and whether the same magnitudes of genetic and environmental effects operate in the two genders. Given knowledge of the prevalence rates in men and women, correlations can be estimated by the methods described in Neale and Cardon (1992).

Consider next an "environmental" model that could explain differing prevalence rates between genders (Cloninger et al., 1978). The assumptions of this

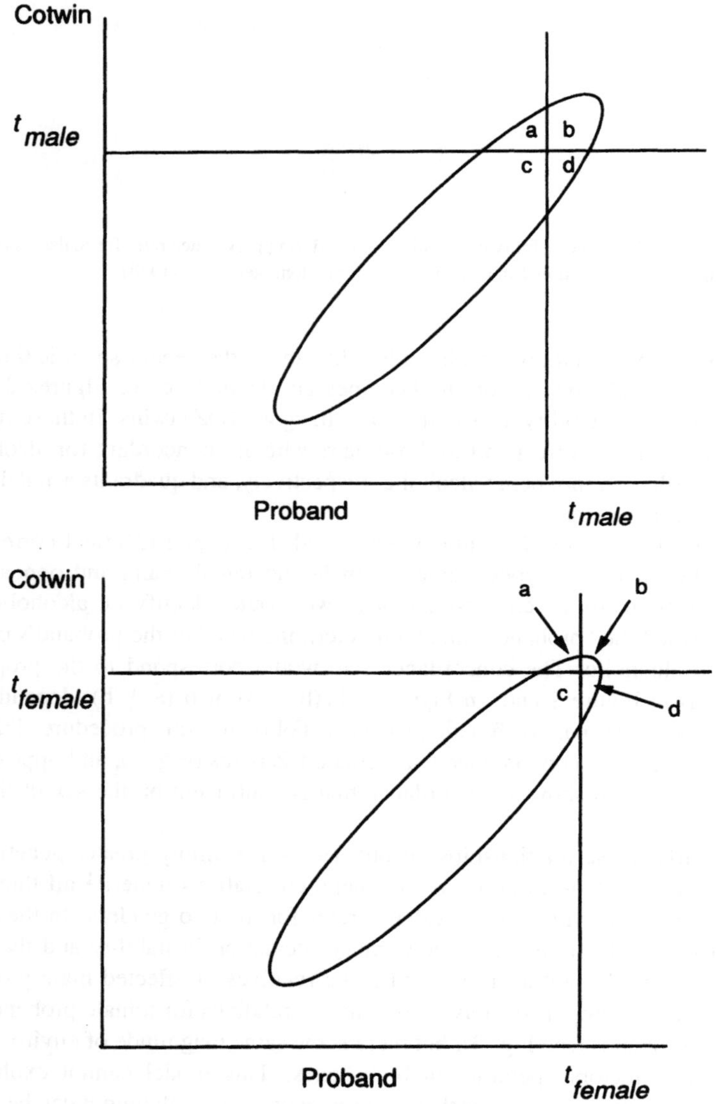

FIGURE 2. Twin concordance for alcoholism under the isocorrelational model: same-sex twins.

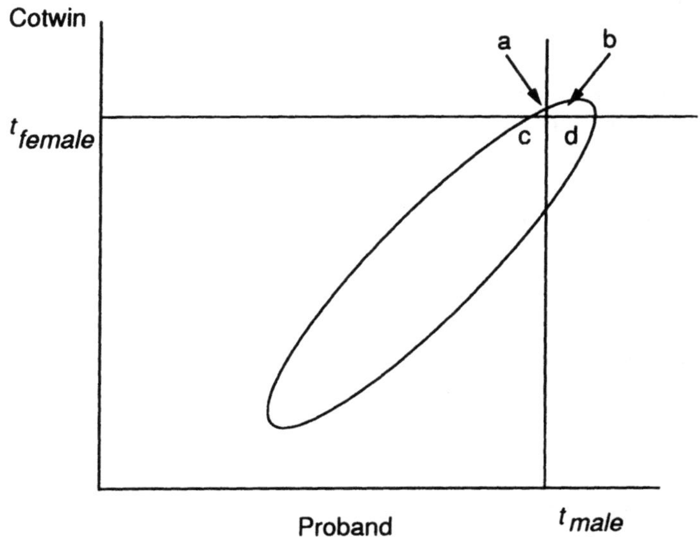

FIGURE 3. Twin concordance for alcoholism under the isocorrelational model: opposite-sex twins.

model are that differences in prevalence rates are due to nonfamilial influences (nonshared environmental influences) that differentially expose the genders to risks or protective influences (note that the exposure is differential between genders and not exclusive to one gender only). Nonshared environmental effects serve to decrease familial resemblance, and therefore the correlation between relatives is lower than it might otherwise be. This model predicts a more similar proportion of affected relatives in female probands and male probands as compared to the isocorrelational model, in which a higher proportion of affected relatives would exist given a female proband than in the case of a male proband (see fig. 3). The familial factors (genetic and *shared* environmental effects) that contribute to variation in liability are assumed to be the same for men and women; therefore, the opposite-sex correlation would be equal to the geometric mean of the same-sex correlations ($r_{mf} = (r_{mm} \cdot r_{ff})^{1/2}$).

The final model presented by Cloninger et al. (1978) was termed the "independent" model. In this model, the familial factors in one gender are imperfectly correlated with the familial factors that influence the other gender. This model has come to be known as the "sex-lmitation" model. Differences in the correlations between relatives of different genders are crucial to understanding whether sex limitation may be present. The difference between this model and the "environmental" model just discussed is that if there is sex limitation of this kind (through differing genetic or environmental mechanisms), then the opposite-sex correlation will be less than the geometric mean of the same-sex correlations. Sex limitation will be discussed in detail shortly, as will the methods for testing for it.

In summary, concordance rates may differ between genders either because there are different degrees of resemblance among male relatives and female relatives or because there are different prevalence rates for men and women. Taking the prevalence rates properly into account will allow us to consider the patterns of resemblance for men and women (e.g., through the computation of tetrachoric correlations for pairs of relatives) and to determine which, if any, of several kinds of sex limitation of the impact of genetic and environmental risk factors are likely to be important.

MEAN GENDER DIFFERENCES

For quantitative traits, such as body mass index (BMI), mean gender differences should be considered in a similar manner to prevalence rates. Often, mean gender differences are observed for a trait where twin and family data suggest the importance of genetic influences to individual variation. The high heritabilities for BMI in both males and females, for example, provide support for the view that individual variability for BMI is substantially influenced by genetic variation (as discussed in Chapters Seven and Eight). However, mean gender differences cannot by extension be attributed definitively to genetic influences. The presence of genetic variation causing individual differences does not necessarily imply that the mean gender difference has a similar etiology, since the underlying etiologies of individual variation have been estimated within each gender.

Though behavior genetic methods are not well positioned to explore the etiology of the mean differences between genders or groups, there have in fact been some attempts to estimate genetic and environmental contributions to group differences (Dolan, Molenaar, & Boomsma, 1992; Turkheimer, 1992). Though they are available, these new methods for testing differential genetic and environmental contributions to mean differences have been little applied to observed data, and their utility has not yet been established.

GENDER DIFFERENCES IN INDIVIDUAL VARIATION

Sex limitation refers to gender-specific effects of genetic or environmental influences. The genes responsible are genes, autosomal or nonautosomal, that are found in *both* genders but are gender-specific in their expression. Thus, unlike any other group comparison (e.g., racial comparisons), there can be no differences in autosomal allele or genotype frequencies, barring postconception selection, though there may be differences in *gene expression*. The most obvious examples of sex-limited genetic expression would be related to primary and secondary sex characteristics, with the expression of these characteristics being dependent on sex hormones (Winchester, 1975). In terms of environmental influences, however, environments can differ both in the magnitude of influences and in kind between males and females.

The extent and nature of sex limitation can be assessed through the use of appropriately designed twin, family, and adoption studies. As the estimation of genetic and environmental components of phenotypic variance is more straightforward for twins than for nontwin families, a most promising test of sex limitation, either genetic or environmental in nature, is to compare twin similarity across same-sex monozytotic (MZ) and DZ pairs and opposite-sex DZ pairs. The information provided by DZ opposite-sex pairs is crucial to testing for sex limitation. If genetic (or environmental) influences differ in kind for men and women, the DZ opposite-sex correlation will be less than the (geometric) average of the two same-sex zygosity correlations. If this difference is not found, there still may be gender differences in the *magnitude* of genetic and environmental variances, though the genetic and environmental influences themselves are assumed to be the same. Alternatively, there may be differences in phenotypic variance, even though the same relative magnitudes of genetic and environmental effects are found in males and females.

This same logic can be extended to the examination of multivariate associations between outcome measures (e.g., blood pressure) and other measured risk factors (e.g., weight or body fat). Thus, an appropriately designed twin study can provide considerable insight into the causes of gender differences in blood pressure, BMI, and alcohol consumption, for instance, and it can pinpoint those aspects of variation and covariation that are most likely to yield to genetic or environmental analysis. The approach of choice to the analysis of the genetic and environmental influences is based on classic path analysis, alternatively called

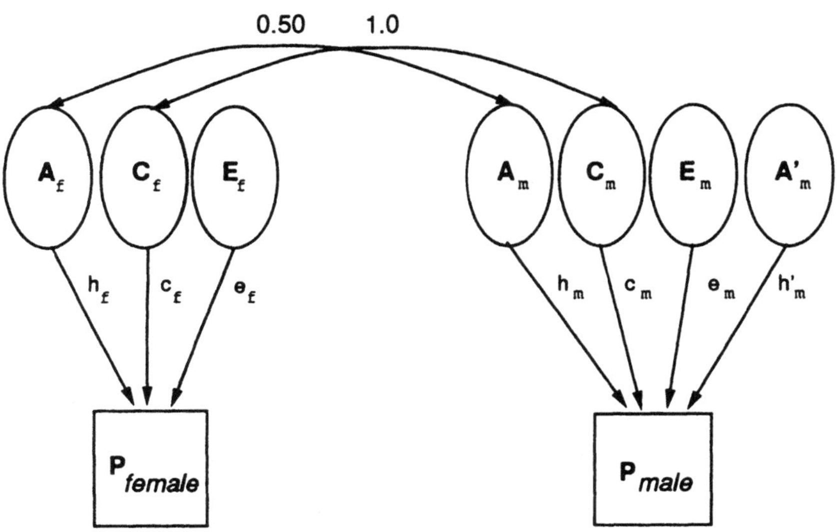

FIGURE 4. One possible genotype–sex interaction model depicting opposite-sex DZ twin pairs and sex limitation of genetic influences.

"linear structural equation modeling" (Heath, Neale, Hewitt, Eaves, & Fulker, 1989; Hewitt, Stunkard, Carroll, Sims, & Turner, 1991; Neale & Cardon, 1992), though there are other widely used methods such as the DeFries–Fulker regression method (DeFries & Fulker, 1985, 1988).

Path analysis formalizes the various possibilities we have discussed in linear path models that lead to quantitative expectations for the observed phenotypic variances and family covariances for twins of each sex and zygosity group. Such a path model is shown in Fig. 4, representing one possible genotype–sex interaction model (Neale & Cardon, 1992) that, in this case, allows for sex limitation of genetic influences. It assumes that to some extent, differing genetic influences are at play in the two genders. One can also modify this model to test for differing familial environmental effects, although with twin data alone one cannot simultaneously explore sex limitation of both genetic and familial environmental effects. To do so, we need other family groups, such as adoptees. However, given the appropriate relatives, the extension of the analysis is straightforward. In Figure 4, the symbols A, C, and E refer to possible sources of influence on the phenotype from, respectively, additive genetic variance, environmental variance shared by opposite-sex twins, and environmental variance not shared by opposite-sex twins. Factors specific to males and not females are denoted by a prime (in this case, A′, a specific additive genetic factor important to males and not females). The model also allows for different magnitudes of genetic and environmental effects across sex. Implicit in the diagram is that the sources of variation and also the magnitude of their impact (indicated by h, c, and e, respectively) may be assumed either to be different for men and women (as shown by the subscripts m and f for male and female, respectively) or to be the same (by equating them).

There are two other submodels of the genotype–sex interaction model: the common-effects model and the scalar-effects model. The common-effects model assumes that the same genetic and environmental effects are present in males and females, but it allows the magnitudes of those genetic and environmental components to differ. The second submodel, the scalar-effects model, assumes the same relative magnitude of genetic and environmental effects, but at the phenotypic level, a scalar multiplier can account for differences in phenotypic variance. This second submodel can be tested only if variance information is available; i.e., it cannot be tested with correlations alone. Finally, one can test these various sex-limitation models against a model that constrains the magnitude of environmental and genetic influences and the overall phenotypic variances to be equal across the sexes [cf. the Cloninger et al. (1978) "isocorrelational" model]. Comparison of the goodness of fit to the observed data under the full sex-limitation model and the alternative submodels provides initial tests for the sex limitation of gene expression and environmental influences. Following such analysis, maximum-likelihood procedures can be used to estimate, for example, the magnitude of the genetic and environmental contributions to the phenotypic variance. Tests of the statistical significance of individual parameters in the model (e.g., genetic influences on the phenotype) can be achieved by the sequential fitting of hierarchically nested models and the use of the difference χ^2 test. These model-fitting and estimation procedures can be implemented using programs such as LISREL (Jöreskog & Sörbom, 1989) or the more recently developed Mx (Neale, 1993), a program specifically developed for analyzing genetically informative twin and family data.

Sex limitation has been tested for as described above in the univariate case for BMI (see Neale & Cardon, 1992). Results and analyses were reported for univariate sex-limitation models fit to BMI data collected on twins in the Virginia Twin Registries (described in Heath et al., 1993; Truett et al., 1994). The DZ opposite-sex correlation was smaller than that for same-sex pairs, suggesting some sex limitation of either genetic or environmental influences. Additionally, females exhibited a greater phenotypic variance for BMI than did males. The genotype–sex interaction model was fit to the data as well as the two submodels, the common- and scalar-effects models. Results indicated that the common-effects model best described the observed pattern of twin correlations, although females exhibited greater genetic and environmental variance than males. However, the relative magnitudes of overall genetic and environmental effects were similar in magnitude: Broad heritability estimates were higher in females (0.75) than in males (0.69), although the difference was not large. Additionally, however, the lower opposite-sex DZ twin correlation was accounted for by nonadditive genetic variation exhibited in males but not in females. Thus, in this case, the twin data suggested that individual differences in BMI among men and among women are determined, in large part, by similar sets of genes, although the manner and magnitude of the expression of these genes may be sex-limited. In particular, we would hypothesize that the impact of any given risk factor would likely be greater in women than in men.

In a similar analysis of twin data for smoking initiation, based on surveys in the United States and Australia, Heath et al. (1993) found that the relative magnitudes of genetic and shared environmental influences were markedly different for men and women, with the additional complication that the effects were culturally dependent. In combined United States samples, shared environmental effects were more important for women than for men, while genetic influences were of greater importance for men than for women. Moreover, there was a strong suggestion that the shared environmental influences themselves were not the same for men and women, "consistent with the possibility that one important influence on smoking initiation is that of same-sex peers, or of similar variables that are more likely to be shared by same-sex than opposite-sex siblings" (Heath et al., 1993, p. 239). Thus, in this case, the analysis of sex limitation points the way to some aspects of gender-specific environmental risks for smoking initiation, both in magnitude and in kind. Furthermore, although we have not described the results in detail, the Heath et al. (1993) report also provides evidence that genetic and environmental risks, and gender differences in their impact, may be dependent on the cultural context in which they are expressed.

The models that have been described are not limited to twin studies; they are equally applicable to other etiologically informative designs such as the adoption design. The twin design is especially powerful in detecting genetic influences, and thus in detecting sex-limited genetic influences, given the inclusion of opposite-sex twin pairs in the analyses. However, greater power to distinguish shared familial environmental influences from genetic influences is found in adoption studies. Thus, a more powerful design to determine sex-limited environmental effects is to compare the resemblance of adopted siblings, both same-sex and opposite-sex, in a manner similar to our illustrative application to twin data. In designs that include parents, one can also compare genetic and environmental transmission between the sexes. It should be noted here, however, that the interpretation of findings could be complicated by genotype–age interactions, as may happen if there is some developmental change in the nature or expression of genetic influences (Hewitt, 1990).

FINAL CONSIDERATIONS AND CONCLUSIONS

In this brief discussion, we have made four main points that we consider to be important in the application of behavior genetics to behavioral medicine. First, average gender differences (in prevalence rates or means) present a prima facie case for considering the possibility that genetic and environmental influences might differentially affect men and women. However, the heritability or environmental etiology of individual differences among men and among women does not lead directly to conclusions about the causes of the average gender differences.

Second, average gender differences, particularly for the prevalence of disorders or categorical conditions (e.g., diagnosis of alcoholism), must be taken into account when interpreting behavior genetic family studies. This fact has been properly appreciated for a long time by behavior and psychiatric geneticists (e.g., Cloninger et al., 1978), but has not always been more widely appreciated.

Third, sex-limited expression of genetic and environmental influences on risk factors may be formally analyzed into differences in the magnitudes of influences and differences in the sources of influence. This distinction will have implications for the study of gender differences in risk factors in behavioral medicine. It is these kinds of gender limitations on the etiology of individual differences *among* men and *among* women that will be most immediately amenable to study.

Finally, the appropriate analysis of gender differences in behavioral medicine, of the kind we have advocated, will require our research designs not only to include both males and females (a practice that has only recently become the norm) but also to ensure the inclusion of *both* same-sex *and* opposite-sex family pairings. Only in this way will we begin to accumulate the kinds of data that will help us formulate a realistic research agenda for the study of gender differences in the operation of genetic and environmental risk factors in behavioral medicine.

REFERENCES

Cloninger, R., Christiansen, K. O., Reich, T., Gottesman, I. I. (1978). Implication of sex differences in prevalences of antisocial personality, alcholism and criminality for familial transmission. *Archives of General Psychiatry, 35,* 941–951.

DeFries, J. C., & Fulker, D. W. (1985). Multiple regression analysis of twin data. *Behavior Genetics, 15,* 467–473.

DeFries, J. C., & Fulker, D. W. (1988). Multiple regression analysis of twin data: Etiology of deviant scores versus individual differences. *Acta Geneticae Medicae et Gemellologiae, 37,* 205–216.

Dolan, C. V., Molenaar, P. C. M., & Boomsma, D. I. (1992). Decomposition of multivariate phenotypic means in multigroup genetic covariance structure analysis. *Behavior Genetics, 22,* 319–335.

Heath, A. C., Cates, R., Martin, N. G., Meyer, J., Hewitt, J. K., Neale, M. C., & Eaves, L. J. (1993). Genetic contribution to risk of smoking across birth cohorts and across cultures. *Journal of Substance Abuse, 5,* 221–246.

Heath, A. C., Neale, M. C., Hewitt, J. K., Eaves, L. J., & Fulker, D. W. (1989). Testing structural equation models for twin data using LISREL. *Behavior Genetics, 19,* 9–36.

Heath, A. C., Slutske, W. S., & Madden, P. A. F. (in press). Gender differences in the genetic contribution to alcoholism risk and drinking patterns. In R. W. Wilsnack and S. C. Wilsnack (Eds.), Gender and alcohol. Rutgers, N.J.: Rutgers University Press.

Hewitt, J. K. (1990). Changes in genetic control during learning, development, and aging. In M. E. Hahn, J. K. Hewitt, N. D. Henderson, & R. Benno (Eds.), *Developmental behavior genetics: Neural, biometrical, and evolutionary approaches* (pp. 217–235). New York: Oxford University Press.

Hewitt, J. K., Stunkard, A. J., Carroll, D., Sims, J., & Turner, J. R. (1991). A twin study approach towards understanding genetic contributions to body size and metabolic rate. *Acta Geneticae Medicae et Gemellologiae, 40,* 133–146.

Jöreskog, K. G., & Sörbom, D. (1989). *LISREL 7: A guide to the program and applications,* 2nd ed. Chicago: SPSS Inc.

Neale, M. C. (1993). *Mx: Statistical modeling* (2nd Ed.). Department of Psychiatry, Medical College of Virginia, Virginia Commonwealth University, Richmond, VA.

Neale, M. C., & Cardon, L. R. (1992). *Methodology for genetic studies of twins and families.* Boston: Kluwer Academic Publishers.

Truett, K. R., Eaves, L. J., Walters, E. E., Heath, A. C., Hewitt, J. K., Meyer, J. M., Silberg, J., Neale, M. C., Martin, N. G., & Kendler, K. S. (1994). A model system for analysis of family resemblance in extended kinships of twins. *Behavior Genetics, 24(1),* 35–49.

Turkheimer, E. (1991). Individual and group differences in adoption studies. *Psychological Bulletin, 110,* 392–405.

Winchester, A. M. (1975). *Human genetics.* Columbus, OH: Charles E. Merrill.

CHAPTER ELEVEN

Studying Ethnicity and Behavioral Medicine

A Quantitative Genetic Approach

KEITH E. WHITFIELD AND TONI P. MILES

INTRODUCTION

Nearly 25 years ago, behavior genetics became an unpopular domain of research in response to the paper by Arthur Jensen (1969) that addressed genetic differences between racial groups. There was outrage at the idea that one racial/ethnic group was genetically inferior to another. This research had reverberations in science at two levels. First, doing behavior genetic studies involving people of color was considered taboo or politically incorrect by all but a few researchers [among them Sandra Scarr (e.g., Scarr, 1981, 1988; Scarr & Weinberg, 1976)]. Second, this research had an effect on scientific studies done on minorities in all fields of inquiry. Scientists avoided studying racial/ethnic groups for fear of showing that minorities did not perform as well as whites on tests designed for whites. This concern over finding deficits in the performance of minorities in cross-cultural studies became a wall that blocked the accumulation of knowledge about minorities.

In fact, behavior genetic methodologies are not designed or based on assumptions of inequality among racial or ethnic populations. These methods allow

---

KEITH E. WHITFIELD • Department of Biobehavioral Health, College of Health and Human Development, Pennsylvania State University, University Park, Pennsylvania 16802. TONI P. MILES • Center for Special Populations and Health, College of Health and Human Development, Pennsylvania State University, University Park, Pennsylvania 16802.

Behavior Genetic Approaches in Behavioral Medicine, edited by J. Rick Turner, Lon R. Cardon, and John K. Hewitt. Plenum Press, New York, 1995.

scientists to identify proportions of variation due to genetic and environmental influences on phenotypes, including behaviors. As scientists are beginning to argue convincingly that within-race studies must be performed prior to conducting cross-cultural studies (Burton, Dilworth-Anderson, & Bengtson, 1992), researchers can benefit by being open to the potential usefulness and power that behavior genetic techniques have for answering questions about minority populations.

The goal of this chapter is to discuss the application of behavior genetic methodology to behavioral medicine studies of racial/ethnic populations. We will discuss links between existing knowledge concerning issues in behavioral medicine such as alcohol use, smoking, and stress in minorities and how these topics can be addressed in a behavior genetic framework. In our examples, we will focus on health behaviors that are of particular importance to various minority populations, but will primarily utilize data on African Americans. In addition, we will discuss the interpretation of differential heritability estimates between ethnic groups.

BEHAVIOR GENETICS AND THE STUDY OF ETHNICITY

One of the most significant contributions to behavioral medicine research that behavior genetic approaches can provide is to elucidate the impact of sociocultural influences on individual variability. For purposes of this discussion, we offer a working definition of sociocultural influences. Sociocultural influences are those environmental, contextual, and familial factors that make individuals of various racial and ethnic groups different from one another. These factors also make individuals in these racial/ethnic groups different from one another because of external pressures from the majority culture as well as historical and experiential dimensions common to individuals of a particular culture. Sociocultural influences are the common beliefs, typical ways that people solve problems, laws and regulations, ideologies, and other factors that are shared by a people. Sociocultural influences are also the structural and institutional forces that affect development. These forces affect people differentially by race. In addition, the folkways and mores of a people serve as the behavioral guidelines for development and interactions with others. They also influence people's perceptions of the world. These forces can create very different environments in which minorities live. The sociocultural influences are not uniform for every member of any group of minorities living in the United States. These influences are also particularly important because they can play a significant role in molding and shaping people's intraracial differences from one another.

Of particular interest to the discussion in this chapter is the identification of individual differences in behavior and health among minorities. Although there is a considerable void in our knowledge about the behavior and health interface of minorities, there now appears to be a concerted effort to increase our knowledge about racial/ethnic minorities that is supported by Federal funding agencies. With this interest in enhancing our knowledge of the diversity of the various groups in the United States, we must begin to reexamine our conceptual "starting points" in framing research questions, hypotheses, and theories on ethnically diverse populations. In addition, the use of research methodologies that can serve to enhance our knowledge about these populations must be integrated into this effort regardless of past research that used the same methodologies wrongly to proliferate ideas of inequalities among races.

The current knowledge of minorities was largely developed through cross-ethnic research (see Burton & Bengston, 1982; Jackson, 1985). Initially, social scientists argued that while ethnic minorities may in fact have some distinctive attributes by virtue of their language, life-style, socioeconomic status, and historical experiences in this country, that distinction was best captured in comparative research between groups. Contemporary researchers, however, argue that the cross-ethnic comparisons have several limitations that preclude a full appreciation of that distinctiveness (Markides, Liang, & Jackson, 1990). One limitation in past cross-ethnic studies involves the comparability of measures used. Quite often, measures are developed, refined, and validated within one group and then applied to other groups under the assumption that their validity and reliability are stable across groups. However, there is evidence indicating that measures used for one group may be inappropriate for others (Morgan & Bengston, 1976).

One of the primary limitations of past cross-ethnic comparisons concerns a lack of attention to within-group variability. In much of the previous research, cross-ethnic comparisons tended to homogenize ethnic groups. Moreover, early comparative research often grouped certain ethnic groups together with little appreciation for subgroup differences. For example, a Hispanic group might include Mexican Americans, Latin Americans, and Puerto Ricans, all of whom reflect varying historical cultures and levels of assimilation. Since these individual subgroups are inherently different, by subsuming them under one ethnic umbrella and then comparing them to whites, important distinctions among the subgroups are lost. Of course, this approach also took little or no notice of the variability within each of the groups that arises from sociocultural influences.

Behavior genetic methods will be particularly useful if one begins to study minorities from the perspective that there is significant *heterogeneity* within populations of minorities. Conceptual and methodological discussions of heterogeneity within minority populations are particularly timely given the changing sociodemographic features of ethnic/racial populations. Individual differences within these groups make it difficult to assess the "average" behavior without first knowing what variability exists as well as the sources of variability. Behavior genetic studies can begin to untangle these complexities by decomposing individual variation into genetic and environmental sources. The particular importance this strategy has for the study of minorities is that the element in the human experience that typically accounts for a significant proportion of variability, namely, the sociocultural environment, can be examined within the conceptualization of the heterogeneous nature of minority populations.

Regionality and Chronic Diseases among Minorities

To illustrate one dimension of heterogeneity that has both genetic and environmental sources of variability that are important in the study of biobehavioral aspects of minority populations, we present a brief discussion of regional differences that exist for various chronic diseases.

Scientists have long understood that human beings function to a large degree in small, relatively isolated groups living in discrete geographic regions. In the United States, patterns of intermarriage and migration indicate that 75–80% of persons were born or married within their place of current residence (Hetzel & Cappetta, 1971). This might suggest that one underlying factor in geographic variation in disease is microvariation in genetic susceptibility. For example, mortality rates from cancer (Pickle, Mason, & Freumeni, 1989), heart disease (Fabsitz & Feinleib, 1980), and stroke (Lanska, 1993) are not uniformly distributed across the United States. Hip fracture incidence, another source of late-life morbidity, is markedly higher in the southern United States (Jacobsen et al., 1990). Analyses of these patterns generally focus on underlying distribution of environmental factors, because these data are also available in the National Center for Health Statistics Surveys.

Most evaluations of the geographic variation in patterns of mortality and morbidity focus almost exclusively on behavioral risk factors such as diet, physical activity, smoking, body weight, and the presence of other comorbid conditions. In addition to individual risk factors, a source of variation in the geographic distribution of disease has been linked to patterns of employment (Foggin & Godon, 1986). In the case of stroke, systematic attempts to investigate the geographic variation in stroke risk factors as an explanation of the geographic variation in stroke frequency and morality have been few (Nefzger, Kuller, & Lillienfeld, 1977) and the results inconsistent (Stoiley, Kuller, & Nefzger, 1977). A question that remains unresolved in these studies can be stated as follows: Why do risk factors such as a high-fat diet or smoking in some regions lead to increased cardiovascular disease, while in others these factors are more closely linked to cancer? In other words, how do the same behavioral factors lead to different chronic disease patterns? An excellent starting point is the study of these patterns using a behavior genetic framework. It seems imperative that one must first identify the sources of the variability in the patterns of chronic disease, and behavior genetic methods can provide this important first step.

This having been said, however, it is important to understand that these behavior genetic methods by themselves are not sufficient for untangling the web of influences that are experienced by America's minority populations. Additional contextual, sociological, and behavioral approaches in research studies will assist in providing a more comprehensive portrait of the human condition that minorities experience.

APPLICATIONS IN THE STUDY OF BEHAVIORAL MEDICINE

There are three primary designs employed in behavior genetic research: classic twin study, parent–offspring family study, and an adoption design (see Chapter 1). Each of these designs is appropriate for the study of individual differences in minority populations. In this section, we will provide a brief overview of three topics of interest to behavioral medicine researchers: alcohol use, smoking, and cardiovascular reactivity to stress. We will also review behavior genetic studies for each topic and conclude each section with a discussion of the insights to be gained from using one of the three research designs in each case.

In this discussion, as noted earlier, the focus will be on research involving African Americans. However, the complexities involved in the study of African Americans also exist for other racial/ethnic groups. Consequently, we assume that attempts to increase our understanding of minority groups (other than African American) could also benefit from employing behavior genetic designs to investigate individual variability within those groups.

Alcohol Use

Alcohol use and susceptibility to alcohol-related diseases have complex genetic and environmental components. Studies of factors that increase the risk of alcohol use and injury among African Americans have focused exclusively on nongenetic factors such as religiosity, urban/northern residence, income (Herd, 1985), and Vietnam era military service (Richards, Goldberg, Anderson, & Rodin, 1990). There are no published reports of quantitative genetic studies of alcohol sensitivity among African Americans, blacks from the Caribbean, or populations living in Africa.

The detrimental effects of alcohol use are of particular importance in studies of health among African Americans because studies with Caucasian samples show a cardioprotective association between light or moderate alcohol use and coronary artery disease mortality (Gaziano et al., 1993). The cardiovascular problem of major public health importance for African Americans is hypertension. Alcohol use has been associated with increases in systolic blood pressure. In studies of African Americans (Ford & Cooper, 1991), Japanese (Langer, Criqui, & Reed, 1992), Chinese (Klag et al., 1993), East Indians (Knight et al., 1993), and Caucasians (DeLabry et al., 1992; Friedman, Klatsky, & Fieglaub, 1993), moderate to heavy alcohol consumption has been identified as a factor that increases systolic blood pressure independently of obesity. Elevated systolic blood pressure is one of the primary independent risk factors for stroke among African Americans (Gillum, 1988).

In the United States, from a population level perspective, there is ample evidence that as a cohort passes through middle age, a large number die before age 65 years from causes that can ultimately be related to alcohol abuse. These alcohol-related deaths in midlife leave behind a pool of old-age survivors who are heterogeneous with respect to historical alcohol use and current health status. While the subset of persons who die prematurely from alcohol-related causes may have a high genetic liability to alcoholism (Goodwin, 1985), persons who survive may have a different (i.e., lower) genetic liability. Since the population aged 65 years and older will continue to increase in numbers between now and the next century, an important and largely unresolved issue is alcohol's contribution to disability as either an independent entity or an initiator of chronic conditions in late life.

Evidence exists from studies of Finnish twins (Kaprio et al., 1987), Australian twins (Jardine & Martin, 1984), Swedish twins (Cederlof, Friberg, & Lundman, 1977), and a cohort of United States Caucasian male twins (Swan, Carmelli, Rosenman, Fabsitz, & Christian, 1990) that there are genetic factors that contribute to variations in alcohol consumption (see also Chapter Two). Heritability estimates for alcohol use across these studies average in excess of 30% of the total variance. Older age and social contact between twins appear to drive these estimates in opposite directions. Heritabilities are lower for older twins, while twins who have frequent contact with each other have inflated estimates. However, the extent of genetic and environmental risks for alcohol-related disability has not been established by previous research, and genetic susceptibility to alcohol and its contribution to both premorbid disability and mortality among African Americans is an area in which no data exist.

One immediate example of an extension of behavior research to the study of alcohol use in African Americans would be afforded by a classic twin study, utilizing data from pairs of twins raised together. An advantage of using twins over siblings or other pairs of relatives is that comparisons are made between genetically related individuals who have experienced developmental events during similar time periods. However, given the relatively low rate of twin births (although this rate is higher in African Americans than that in other ethnic groups in the United States), a twin study of African Americans might require a data collection site that is densely populated with African Americans. The alternative to identifying such a population would be to identify twins using a national population-based sample. Such a registry of twins does exist in the Vietnam Era Twin Registry (Eisen, True, Goldberg, Henderson, & Robinette, 1987). This registry is an excellent source for a behavior genetic study of alcohol use among adult African Americans, although it is limited in that it includes only men.

There are clearly gaps in our current knowledge concerning the contribution made by genetic factors to alcohol use and the influence these factors have on physical health in adulthood among African Americans. To be able to design public health interventions that will result in reduced alcohol-related morbidity and mortality, we need to study the utility of using end points such as health status, comorbid medical conditions, and functioning. The contribution of inheritance to alcohol-induced disease may also be more sharply defined when contextual variables such as geographic residence and social contacts are also well characterized.

SMOKING

As a risk factor, smoking is one of the primary behaviors that contribute to the difference in mortality rates between Caucasians and African Americans. There is a strong relationship between smoking and cardiovascular disease and some forms of cancer (e.g., lung, cervical, oral, esophageal, pancreatic, renal, and bladder cancers) (American Cancer Society, 1992). African Americans experience higher age-adjusted morbidity and mortality rates of both coronary heart disease and stroke (NHLBI, 1985). These higher rates are due in part to the increased prevalence of smoking noted in African Americans. The prevalence of smoking in African Americans is 34% as compared to 29% in Caucasians (Pierce, Fiore, Novotny, Hatziandreu, & Davis, 1989). These differences suggest that there might be cross-cultural differences in the relative impact that familial environment and genes have on current smoking status. In their analysis of the available twin data on risk of being a current smoker, Heath and Madden (see Chapter 3) found differences in estimates between studies at the same time period using similar assessment protocols but samples from different countries (e.g., Finnish, Swedish, and American) (see Chapter 3 for details).

Although only a few behavior genetic studies have focused on the heritability of smoking behavior, the available literature does support the existence of genetic influences. A review of the current behavior genetic literature on smoking behavior as well as calculations of genetic and environmental influences are presented in Chapter Three. In that chapter, the authors calculated that 55% of the variance associated with the risk of being a current smoker in women and 56% in men was due to genetic sources. They also found that 24% of the variance in women and 22% in men was due to shared environmental effects, and that within-family environmental variation accounted for the remaining 21% in women and 22% in men.

The results of these previous studies provide an intriguing question to ask about African Americans: What are the proportions of genetic and environmental influences in African Americans? We expect that the proportions of genetic and environmental influences would be different from those in a comparable sample of Caucasian Americans. We hypothesize that there will be greater proportions of environmental influence (shared and within-family variation). As a recent example, consider the environmental "pressure" exerted by the targeting of cigarette advertisements in regions that are densely populated with African Americans.

An adoption study designed to investigate smoking behavior would be another example of an appropriate application of behavior genetic methodology in a sample of African Americans, since the adoption design provides the strongest assessment of environmental influences. However, such studies are the most difficult to conduct because the constellation of parents is twice that necessary for a familial study: Both adopted and biological parents are assessed. One important aspect of African American communities and families is that there are frequently informal adoptions (Taylor, 1986; Taylor, Chatters, Tucker, & Lewis, 1990). For example, friends will frequently assume the care of a child because the family of origin has severe financial difficulties (Taylor et al., 1990). In addition, there are scenarios in which other family members, such as grandparents, will take the responsibility of a child born out of wedlock (Burton, 1992; Tinsley & Parke, 1984). In the latter case, ascertaining the genetic influence must take account of genetic similarity between the grandparents and grandchildren (the grandparent–grandchild additive genetic correlation is 0.25).

We pose the possibility of an intergenerational adoption design as one of the ways in which knowledge of African American families can be an avenue to extend behavior genetic designs and provide an approach to the study of familial influences on traits. Attempting to conduct such a study would of course raise a number of specific methodological issues, such as assessing the amount of contact that the child has with the biological parents.

Cardiovascular Reactivity to Stress

The number one health problem in African Americans is hypertension, with a 75–80% prevalence by age 65 (National Center for Health Statistics, 1991). African Americans experience higher age-adjusted morbidity and mortality rates of both coronary heart disease and stroke (NHLBI, 1985). These higher rates are due in part to the increased prevalence of hypertension noted in African Americans.

There is currently great interest in the role of cardiovascular reactivity in the development of hypertension (Anderson, McNeilly, Armstead, Clark, & Pieper, 1993). Cardiovascular reactivity has been defined as the "changes in physiological parameters that occur when individuals are exposed to behavioral, psychological, or physical challenges" (Anderson et al., 1993). There is evidence to suggest that African Americans frequently demonstrate higher peripheral vascular reactivity than Caucasians (Anderson, McNeilly, & Myers, 1991). Given the prevalence of hypertension in African Americans, the relationship between autonomic reactivity and hypertensive risk represents an important domain of behavioral medicine inquiry in the study of ethnicity. Investigations into this domain require attention to the individual differences in stress-induced cardiovascular responses.

Behavior genetic approaches have been used in the study of stress responses. We refer the reader to Chapter 5, which provides a review of pertinent literature on the topic. A summary of the review of relevant literature in Chapter 5 reveals that (1) heart rate and blood pressure responses to laboratory psychological stressors are moderately heritable and (2) there are probably no shared environmental effects involved in individual differences in response to stress. Of the studies presented in Chapter 5, none systematically included minorities in its sample. Anderson, McNeilly, and Myers (1992) have noted that African Americans, on average, are exposed to a wider array of chronic stressors than Caucasians. This wider array of chronic stressors might be one of the driving forces in differences in the proportion of genetic and environmental sources of individual differences found when ethnicity is included as a variable of interest.

Employing a family design is yet another behavior genetic approach that might be used to study African Americans. A family design provides upper estimates of heritability (genetic and shared environmental influences) as well as an estimate of nonshared environmental influences. However, in carrying out family studies with an African American population, one must be mindful of the diversity within and across families, as there has historically been greater variability in the family structure of this group than research has previously suggested (Dilworth-Anderson, Burton, & Boulin-Johnson, 1993).

In terms of diversity of intergenerational family structure among African Americans, Bengston, Rosenthal, and Burton (1990) have identified six family configurations. The first is the "beanpole" family. Also called the "verticalized" family, this structure is characterized by intergernerational extension—meaning that there are three or more living generations in a family—and intragenerational contraction—meaning that there are declining numbers of members within each generation. The age-condensed family structure is the second configuration and is characterized by multiple living generations with a relatively short age span between generations. The third is the age-gapped, defined as having multiple generations with large age distances between tiers. The fourth reflects the one-generation family structure of the childless family. A matrilineal intergenerational family structure represents the fifth configuration. And sixth, a stepfamily model or "blended model" is characterized by the merging of families from two or more previous marriages. This is a family configuration in which increasingly more elderly people will live their lives.

The diversity of African American families presents several possibilities for extending traditional behavior genetic approaches to studying the family that might provide useful insights into the study of cardiovascular reactivity, as well as other phenotypes of concern in behavioral medicine.

DIFFERENTIAL HERITABILITY

One of the potentially most disturbing results that might be produced from a behavior genetic study of health-related variables is that a minority group shows a lower heritability for a desirable phenotype or higher heritability for undesirable characteristics than their white counterparts. The danger here is the possible interpretation that the majority population may have a "superior set of genes" because they show a higher heritability on positive measures of health. This interpretation is *erroneous* given the assumptions of behavior genetic methodology. A more plausible interpretation of differential heritability between two ethnic groups is that the expression of the genetic effects is different between the two groups. The expression of the phenotype is the result of genetic *and* environmental factors working in concert. Variations in the genetic makeup of various ethnic minority groups are so small that, genetically, they are essentially the same as in the majority culture (Black, 1992). The expressed phenotype, then, is different because the environment that acts in concert with, and in addition to, genetic expression is different. Thus, the sociocultural influence of different environments produces different proportions of genetic and environmental influences, not the identification of superior or inferior genetic constitutions. In the study of variables of interest in behavioral medicine, one should expect to find different heritabilities across different groups, given that minority populations often live in poor environments filled with risk factors.

CONCLUSIONS

The challenge for behavioral medicine researchers is to avoid relying on dated typologies of ethnic minority populations and to begin to focus on the variability that exists within these groups. An appreciation of diversity assists in fully understanding biobehavioral aspects of development of all peoples. The issue of diversity is currently at the forefront of much of science. A behavior genetic perspective is an excellent starting point for understanding the individual differences that exist for minority populations.

Because of the social and environmental heterogeneity within minority groups, the source, recruitment, and retention of participants in our studies can be significant sources of unexpected bias. Careful attention to the ascertainment of samples is required to avoid overestimation or underestimation of genetic and environmental effects. There are some existing nongenetic studies of the possible hazards and the sampling methodologies designed to avoid those pitfalls (see Herzog & Rodgers, 1982; Jackson, 1989).

Studies of genetic and environmental components that contribute to the health of minority populations must be undertaken with clarity of purpose. Behavior genetic models are powerful tools. But, used naïvely, they can produce results that can be misunderstood. It is such misapplication that creates chaos in attempting to build knowledge bases about minorities. The concepts and design issues discussed in this chapter are presented as starting points for a path of research that has important and far-reaching implications for policy and public health practices. It is our hope that these designs will be useful in spawning research in this area of study.

REFERENCES

American Cancer Society (1992). *Cancer facts and figures*. Atlanta: American Cancer Society.

Anderson, N. B., McNeilly, M. D., Armstead, C., Clark, R., & Pieper, C. (1993). Assessment of cardiovascular reactivity: A methodological review. *Ethnicity and Disease, 3,* S29–S37.

Anderson, N. B., McNeilly, M. D., & Myers, H. (1991). Autonomic reactivity and hypertension in blacks: A review and proposed model. *Ethnicity and Disease, 1,* 154–170.

Anderson, N. B., McNeilly, M. D., & Myers, H. (1992). Toward understanding race differences in autonomic reactivity: A proposed contextual model. In J. R. Turner, A. Sherwood, & K. C. Light (Eds.), *Individual differences in cardiovascular response to stress* (pp. 125–145). New York: Plenum Press.

Bengston, V. L., Rosenthal, C., & Burton, L. M. (1990). Families and aging: Diversity and heterogeneity. In R. H. Binstock & L. K. George (Eds.), *Handbook of aging and the social sciences* (pp. 263–287). New York: Academic Press.

Black, F. L. (1992). Why did they die? *Science, 258,* 1739–1740.

Burton, L. M. (1992). Black grandparents rearing children of drug addicted parents: Stressors, outcomes, and social service needs. *Gerontologist, 32(6),* 744–751.

Burton, L. M., & Bengston, V. L. (1982). Research in elderly minority communities: Problems and potentials. In R. C. Manuel (Ed.), *Minority aging: Sociological and social psychological issues* (pp. 215–222). Westport, CN: Greenwood.

Burton, L. M., Dilworth-Anderson, P., & Bengston, V. L. (1992). Creating new ways of thinking about diversity and aging: Theoretical challenges for the twenty-first century. *Generations, 15(4),* 67–72.

Cederlof, R., Friberg, L., & Lundman, T. (1977). The interactions of smoking, environment and heredity and their implications for disease etiology: A report of epidemiological studies on the Swedish Twin Registries. *Acta Medica Scandanavica, 202(Supplementum 612),* 1–128.

DeLabry, L. O., Glynn, R. J., Levenson, M. R., Hermos, J. A., LoCastro, J. S., & Vokonas, P. (1992). Alcohol consumption and mortality in an American male population. *Journal on the Study of Alcohol, 53,* 25–32.

Dilworth-Anderson, P., Burton, L. M., & Boulin-Johnson, L. (1993). Reframing theories for understanding race, ethnicity, and family. In P. Boss, W. Doherty, R. Larossa, W. Schumm, & S. Steinmetz (Eds.), *Source book of family, theories, and methods: A contextual approach* (pp. 627–646). New York: Plenum.

Eisen, S., True, W., Goldberg, J., Henderson, W., & Robinette, D. (1987). The Vietnam Era Twin (VET) Registry: Method of construction. *Acta Geneticae Medicae et Gemellalogiae, 36,* 61–66.

Fabsitz, R., & Feinleib, M. (1980). Geographic patterns in county mortality rates from cardiovascular diseases. *American Journal of Epidemiology, 111(3),* 315–328.

Foggin, P., & Godon, D. (1986). Cardiovascular mortality as it relates to the geographic distribution of employment in nonmetropolitan Quebec. *Social Science and Medicine, 22(5),* 559–569.

Ford, E. S., & Cooper, R. S. (1991). Risk factors for hypertension in a national cohort study. *Hypertension, 18,* 598–606.

Friedman, G. D., Klatsky, A. L., & Fieglaub, A. B. (1993). Alcohol intake and hypertension. *Annals of Internal Medicine, 98*[(5) Part 2], 846–849.

Gaziano, J. M., Buring, J., Breslow, J. L., Goldhaber, S. Z., Rosner, B., VanDenburgh, M., Willett, W., & Hennekens, C. H. (1993). Moderate alcohol intake, increased levels of high-density lipoprotein and its subfractions, and decreased risk of myocardial infarction. *New England Journal of Medicine, 329(25),* 1829–1834.

Gillum, R. F. (1988). Stroke in blacks. *Stroke, 19,* 1–9.

Goodwin, D. W. (1985). Alcoholism and genetics: The sins of the fathers. *Archives of General Psychiatry, 42,* 171–174.

Herd, D. (1985). The epidemiology of drinking patterns and alcohol-related problems among U.S. blacks. In: Proceedings of a conference on the Epidemiology of Alcohol Use and Abuse among ethnic minority groups, September 1985. NIAAA Research Monograph No. 18.

Herzog, A. R., & Rodgers, W. L. (1982). *Surveys of older Americans: Some methodological investigations: Final Report to the National Institute on Aging.* Ann Arbor: University of Michigan, Institute for Social Research.

Hetzel, A. M., & Cappetta, M. (1971). *Marriage: Trends and characteristics—United States.*

USPHS Publication No. 72—1007, Series 21, No. 21. Hyattsville, MD: National Center for Health Statistics.

Jackson, J. J. (1985). Race, national origin, ethnicity, and aging. In R. H. Binstock & E. Shanas (Eds.), *Handbook of aging and social sciences* (pp. 264–303). New York: Van Nostrand-Reinhold.

Jackson, J. (1989). Methodological issues in survey research on older minority adults. In M. Powell Lawton & A. Regula Herzog (Eds.), *Special research methods for gerontology.* Amityville, NY: Baywood.

Jacobsen, S., Goldberg, J., Miles, T., Brody, J., Steirs, W., & Rimm, A. (1990). Regional variation in the incidence of hip fracture among white females aged 65 years and older in the United States. *Journal of the American Medical Association, 263,* 500–502.

Jardine, R., & Martin, N. (1984). Causes of variation in drinking habits in a large twin sample. *Acta Geneticae Medicae et Gemellalogiae, 33,* 435–450.

Jensen, A. R. (1969). How much can we boost IQ and scholastic achievement? *Harvard Educational Review, 39,* 1–123.

Kaprio, J., Koskenvuo, M., Langinvainio, H., Romanov, K., Sama, S., & Rose, R. (1987). Genetic influences on use and abuse of alcohol: A study of 5638 adult Finnish twin brothers. *Alcoholism: Clinical and Experimental Research, 11(4),* 349–356.

Klag, M. J., He, J., Whelton, P,. Chen, J. Y., Qian, M. C., & He, G. Q. (1993). Alcohol use and blood pressure in an unacculturated society. *Hypertension, 22,* 365–370.

Knight, T., Smith, Z., Lockton, J. A., Fahota, P., Bedford, A., Toop, M., Kernohan, E., & Baker, M. R. (1993). Ethnic differences in risk markers for heart disease in Bradford and implications for preventive strategies. *Journal of Epidemiology and Community Health, 47,* 89–95.

Langer, R. D., Criqui, M., Reed, D. M. (1992). Lipoproteins and blood pressure as biological pathways for effect of moderate alcohol consumption on coronary heart disease. *Circulation, 85,* 910–915.

Lanska, D. J. (1993). Geographic distribution of stroke mortality in the United States: 1939–1941 to 1979–1981. *Neurology, 43,* 1839–1851.

Markides, K. S., Liang, J., & Jackson, J. S. (1990). Race, ethnicity, and aging: Conceptual and methodological issues. *Handbook of aging and social sciences* (pp. 112–129). New York: Van Nostrand-Reinhold.

Morgan, L. A., & Bengston, V. L. (1976). Measuring perceptions of aging across social strata. Paper presented at the 29th annual meeting of the Gerontological Society, New York, October 13–17, 1976.

National Center for Health Statistics (1991). *Health in the United States, 1990.* Hyattsville, MD: U.S. Public Health Service.

Nefzger, M. D., Kuller, L. H., & Lillienfeld, A. M. (1977). Three-area epidemiologic study of geographic differences in stroke mortality. I. Background and methods. *Stroke, 8,* 547–550.

National Heart, Lung, and Blood Institute (1985). Hypertension prevalence and the status of awareness treatment and control in the U.S.: Final report of the subcommittee on definition and prevalence of the 1984 Joint National Committee. *Hypertension, 7,* 457–468.

Pickle, L. W., Mason, T. J., & Freumeni, J. F. (1989). The new United States Cancer Atlas. *Recent Results in Cancer Research, 114,* 196-207.

Pierce, J. P., Fiore, M. C., Novotny, T. E., Hatziandreu, E. J., & Davis, R. M. (1989). Trends in cigarette smoking in the U.S. *Journal of the American Medical Association, 261,* 56–60.

Richards, M. S., Goldberg, J., Anderson, R. J., & Rodin, M. B. (1990). Alcohol consumption and problem drinking in Vietnam era veterans and nonveterans. *Journal on the Study of Alcohol, 51,* 396–402.

Scarr, S. (1981). *Race, social class and individual differences in IQ.* Hillsdale, NJ: Erlbaum Associates.

Scarr, S. (1988). Race and gender as psychological variables: Social and ethical issues. *American Psychologist, 43,* 56-59.

Scarr, S., & Weinberg, R. A. (1976). IQ test performance of black children adopted by white families. *American Psychologist, 31,* 726-739.

Stoiley, P. D., Kuller, L. H., & Nefzger, M. D. (1977). Three-area epidemiologic study of geographic differences in stroke mortality. II. Results. *Stroke, 8,* 551–557.

Swan, G. E., Carmelli, D., Rosenman, R. H., Fabsitz, R. H., & Christian, J. (1990). Smoking and alcohol consumption in adult male twins: Genetic heritability and shared environmental influences. *Journal of Substance Abuse, 2,* 39–50.

Taylor, R. J. (1986). Aging and supportive relationships among black Americans. In J. S. Jackson (Ed.), *The black american elderly* (pp. 259–281). New York: Springer.

Taylor, R. J., Chatters, L. M., Tucker, M. B., & Lewis, E. (1990). Developments in research on black families: A decade review. *Journal of Marriage and the Family, 52,* 993–1014.

Tinsley, B. R., & Parke, R. (1984). Grandparents as support and socialization agents. In M. Lewis (Ed.), *Beyond the dyad* (pp. 161–191). New York: Plenum Press.

PART SIX

BEYOND HERITABILITY

CHAPTER TWELVE

Genetics, Environmental Risks, and Protective Factors

ROBERT PLOMIN

INTRODUCTION

Genetic differences among individuals contribute to most domains of behavior (Plomin & McClearn, 1993) and to many common medical disorders (King, Rotter, & Motulsky, 1992). As previous chapters indicate, genetics also plays a role in constructs central to behavioral medicine, such as alcohol use and abuse, smoking, cardiovascular reactivity, and obesity. More research is needed to investigate whether and how much genetic factors influence other behaviors involved in promoting health and in preventing and treating disease (Plomin & Rende, 1991).

This chapter considers recent advances in quantitative genetics that can be used to take us far beyond the simplest "whether" and "how much" questions of heritability. While the emphasis of the chapter is on the contribution of quantitative genetic analysis to our understanding of *environmental* influences, it begins with other examples: multivariate analysis, developmental analysis, extremes analysis, and quantitative trait loci.

BEYOND HERITABILITY

MULTIVARIATE ANALYSIS

Multivariate analysis assesses genetic contributions to covariance between traits rather than to the variance of each trait considered separately. In this way,

ROBERT PLOMIN • Centre for Social, Genetic, and Developmental Psychiatry, Institute of Psychiatry, London, United Kingdom SE5 8AF.

Behavior Genetic Approaches in Behavioral Medicine, edited by J. Rick Turner, Lon R. Cardon, and John K. Hewitt. Plenum Press, New York, 1995.

multivariate analysis can address the genetic and environmental etiology of relationships between behavior and biology. For example, in associations between personality and disease (e.g., Friedman, 1990), to what extent are genetic factors responsible for the overlap? Multivariate analysis can also broach the fundamental issues of heterogeneity and comorbidity of dimensions and disorders at the level of etiology rather than symptomatology (Hewitt, 1993). For example, what is the structure of genetic influences on the component mechanisms—e.g., changes in peripheral resistance and changes in cardiac output—that underlie cardiovascular reactivity (Turner, 1994)?

Developmental Analysis

Developmental analysis tracks change as well as continuity in genetic effects during development. One type of developmental analysis asks whether the relative influence of genetic and environmental factors changes during development. For example, does heritability decrease during the life course as environmental effects accumulate? Despite the reasonableness of this hypothesis, the available developmental data suggest the opposite: When heritability changes during development, it tends to increase rather than decrease (Plomin, 1986).

The second type of developmental question addresses the genetic and environmental etiology of age-to-age change and continuity using longitudinal data (Eaves, Long, & Heath, 1986; Hewitt, Eaves, Neale, & Meyer, 1988). That is, to what extent do genetic effects at one age overlap with genetic effects at another age? For example, in a longitudinal study of about 4000 pairs of adult twins at 20 and 45 years of age, the heritability of weight and body mass index was about 80% at both ages (Stunkard, Foch, & Hrubec, 1986). However, longitudinal genetic analysis suggested that about 25% of the genetic effects at 45 years differed from genetic effects at 20 years. Developmental changes in genetic effects should not be surprising, because the processes involved in weight and weight gain seem likely to differ from young adulthood to middle adulthood. The general theme of developmental analysis is that heritability should not be equated with stability or immutability.

Extremes Analysis

Extremes analysis addresses genetic links between normal dimensions and abnormal disorders. If, as seems likely, multiple genes are responsible for genetic influences on complex dimensions and disorders, it is reasonable to hypothesize that a continuum of genetic risk extends from normal to abnormal behavior. For example, is alcohol abuse merely the extreme of a continuous dimension of genetic and environmental variability in alcohol use? A quantitative genetics technique developed during the past decade investigates the extent to which a disorder is the etiological extreme of a continuous dimension (DeFries & Fulker,

1985). Preliminary research using this approach suggests that some common disorders represent the genetic extremes of continuous dimensions.

Molecular Genetic Analysis

Molecular genetic analysis that attempts to identify specific genes responsible for complex dimensions and disorders is now possible (Plomin, Owen, & McGuffin, 1994). Because this technique promises to revolutionize genetic research on behavior, it is discussed in somewhat more detail than the other advances that go beyond heritability.

Many rare disorders—such as familial hypercholesterolemia caused by mutations in a low-density-lipoprotein (LDL) receptor (Hobbs, Russell, Brown, & Goldstein, 1990)—show simple Mendelian patterns of inheritance for which defects in a single gene are the necessary and sufficient cause of the disorder. Linkage analysis and the rapidly expanding map of the human genome guarantee that the genes responsible for such single-gene disorders will be mapped and eventually cloned, as has already happened for scores of single-gene disorders. The new frontier for molecular genetics lies with common and complex dimensions, disorders, and diseases central to behavioral medicine. The challenge is to use molecular genetic techniques to identify genes involved in complex systems influenced by multiple genes as well as multiple nongenetic factors, especially when no single gene is either necessary or sufficient.

The traditional reductionistic approach of molecular biology assumes that a single gene is responsible for a disorder, a view that could be called the "one-gene, one-disorder" (OGOD) hypothesis. It is possible that complex traits are genetically simple. For example, it has recently been reported that about 75% of the total genetic effect on bone mass, the major determinant of risk for osteoporotic fracture, can be predicted by alleles of a vitamin D receptor gene (Morrison et al., 1994). However, one should bear in mind H. L. Mencken's dictum: "For every complex question there is a simple answer—and it is wrong." A more contemporaneous variant of the OGOD approach assumes that a complex disorder is actually a concatenation of several disorders, each caused by a different gene. For example, the autosomal dominant LDL-receptor mutations responsible for familial hypercholesterolemia may be one of many OGOD causes of coronary heart disease (CHD) that together account for genetic influence on CHD.

In contrast to the OGOD perspective is the view that genetic influences on complex, common disorders are largely due to multiple genes of varying effect size that contribute additively and interchangeably, as do risk factors to vulnerability (Plomin et al., 1994). Any single gene in such a multigene system is neither necessary nor sufficient to cause the disorder. In other words, genetic effects involve probabilistic propensities rather than predetermined programming. One implication of the assumption of a multigene system is that phenotypes are distributed dimensionally even if traits happen to be assessed using dichotomous diagnoses. For this reason, genes that contribute to multigenic systems have been called *quantitative trait loci* (QTLs) (Gelderman, 1975). The term *QTL* can be distinguished from the term *polygenic,* which simply means multiple genes but has come to connote many genes of small effect size that are *not* individually identifiable. The term QTL is used to denote multiple genes of any effect size, possibly varying from locus to locus, such that they *may* be individually identifiable.

CHD, the principal cause of death in the United States, is emerging as a model for the identification of genes in complex common diseases. The LDL-receptor gene responsible for familial hypercholesterolemia accounts for only a very small proportion of individuals with premature myocardial infarction (Motulsky & Brunzell, 1992). In contrast, QTLs have been identified that account for as much as a quarter of the genetic variance of CHD (Humphries, 1988; Sing & Boerwinkle, 1987). All major lipoprotein genes and many other genes relevant to the hyperlipidemias have been mapped and cloned. Certain alleles of apoprotein-A (apo-A) (involved in high-density-lipoprotein function and formation), apo-B (the principal constituent of LDL cholesterol), and apo-E (related to plasma cholesterol levels) contribute to risk for CHD. Nonlipid genetic factors such as a deletion polymorphism in the angiotensin-converting enzyme (ACE) gene (Cambien et al., 1992) are also associated with risk for CHD.

The most important implication of the distinction between the OGOD and QTL perspectives is that much greater statistical power is needed to detect QTLs because their effects can be much smaller. OGOD effects segregate simply in families, and linked deoxyribonucleic acid (DNA) markers cosegregate with the disorder. However, conventional linkage analysis of large pedigrees is unlikely to have sufficient power to detect QTLs for a disorder unless a particular locus accounts for most of the genetic variance, in which case it is not really a QTL. Moreover, QTLs imply quantitative dimensions, and these are not easily handled by linkage approaches, which traditionally assume a dichotomous diagnosis that cosegregates with a single gene. Newer linkage designs such as affected relative pair designs are more robust than traditional pedigree studies, since they do not depend on assumptions about mode of inheritance (Risch, 1990). It is also possible to incorporate quantitative measures in linkage studies (Fulker & Cardon, 1994). These newer linkage designs may be able to detect QTLs of large effect size.

The advantage of linkage approaches is that they can detect OGOD effects without a priori knowledge of pathological processes in a systematic search of the genome using a few hundred highly polymorphic DNA markers. Such systematic screens of the genome can also exclude OGOD effects. However, they cannot exclude small QTL effects, at least with realistic sample sizes.

Although linkage remains the strategy of choice from the OGOD perspective and for skimming off the largest QTL effects, other strategies are needed to detect QTLs of smaller effect size. Most likely, given the pace of discovery in molecular biology, new techniques will be developed to reach this goal. For the present, allelic association represents an increasingly used strategy that is complementary to linkage (Owen & McGuffin, 1993). In contrast to linkage, allelic association refers to a correlation in the population between a phenotype and a particular allele, usually assessed as an allelic or genotypic frequency difference between cases and controls (Edwards, 1991). Despite its limitations, allelic association can at least provide the statistical power needed to detect QTLs of small effect size. For example, the frequency of individuals homozygous for the ACE deletion was 32% for patients with myocardial infarction and 27% for control subjects. This slightly increased relative risk of 1.3, which accounts for less than 1% of the liability to the disorder, was highly significant statistically because the sample was extremely large (610 cases and 733 controls). The ACE deletion has been shown to be physiologically functional; i.e., it is not just a DNA marker near a functional gene. Use of such functional polymorphisms greatly enhances the power of the allelic association approach to detect QTLs (Sobell, Heston, & Sommer, 1992). The new generation of complementary DNA markers and techniques to detect point mutations in coding sequences are rapidly producing markers of this type.

Harnessing the power of molecular genetics to identify QTLs that contribute genetic risk for a disorder, but are neither necessary nor sufficient to cause the disorder, marks the dawn of a new era in genetic research on complex human dimensions and disorders. This topic is taken up again in Chapter 13.

GENETIC RESEARCH AND IDENTIFICATION OF ENVIRONMENTAL INFLUENCES

Perhaps the most important way in which genetics research has gone beyond merely documenting genetic influence involves the implications of genetic research for understanding environmental influences (Plomin, 1994a).

Nurture as Well as Nature

Genetics research provides the best available evidence for the importance of nonheritable factors. Usually, genetic factors do not account for more than about half the variance for behavioral and medical disorders and dimensions. The current enthusiasm for genetics should not obscure the important contribution of nonheritable factors, even though these factors are more difficult to investigate. For environmental transmission, there is nothing comparable to the laws of hereditary transmission or to the gene as a basic unit of transmission. It should be noted that the "environment" in quantitative genetics denotes all nonheritable factors, including nontransmissible DNA events such as somatic mutation, imprinting, and unstable DNA sequences in addition to the perinatal and psychosocial factors that are usually investigated in environmental research.

Nonshared Environment

Two specific discoveries from genetic research have far-reaching implications for understanding environmental influences. The first is nonshared environment. The way in which the environment influences behavioral development has been shown to contradict socialization theories from Freud onward. For example, the fact that behavioral and medical dimensions and disorders run in families has reasonably but wrongly been interpreted environmentally. Research shows that genetics generally accounts for familial resemblance. The way the environment works for most disorders and dimensions is to make children growing up in the same family different, not similar (Plomin & Daniels, 1987). For this reason, such environmental effects are called "nonshared environment" (or within-family, individual, or specific environment).

Nonshared environment accounts for nearly all environmental variance for personality as assessed by questionnaire, for psychopathology, and, after childhood, for cognitive abilities (Plomin, Chipuer, & Neiderhiser, 1994). Even for attitudes, such as the major attitudinal factor called "traditionalism" (conservatism), environmental influences are primarily due to nonshared environment, once the high assortative mating for such attitudes is taken into account (Eaves, Eysenck, & Martin, 1989). As indicated in Chapter 2, modest concordance of identical twins for alcoholism suggests substantial effects of nonshared environment in the etiology of alcoholism. (Differences within pairs of identical twins can be explained only by nonshared environment and error of measurement.) Another surprising example for behavioral medicine is obesity. Twin and adoption studies converge on the conclusion that environmental influences for weight and obesity are entirely nonshared (Grilo & Pogue-Geile, 1991) (see also Chapters 7 and 8). That is, growing up in the same family does not make individuals similar for weight or obesity despite apparently similar diets.

The message of nonshared environment is not that family experiences are necessarily unimportant, but rather that the relevant environmental influences are specific to each child, not general to an entire family (Dunn & Plomin, 1990). These findings suggest that we need to think about the environment on an individual-by-individual basis. The critical question is: Why are children in the same family so different? The answers will come from research that includes more than one child in each family and identifies experiences that differ between children growing up in the same family.

The second example of genetic research relevant to understanding the environment is the focus of the rest of this chapter.

THE NATURE OF NURTURE

Addictive behaviors show genetic influence, as discussed in preceding chapters. What is more surprising is that *exposure* to drugs shows genetic influence. In a twin study of 1026 male twin pairs from the Vietnam Era Twin

Registry, significant heritability was found, not just for use of marijuana, stimulants, sedatives, cocaine, opiates, and psychedelics, but also for exposure to each drug (Tsuang et al., 1992). Exposure to drugs is usually, and reasonably, considered an environmental risk variable. For this reason, drug use as a result of exposure to drugs is also thought to be due to environmental factors. Contrary to these assumptions, which are so reasonable as to be implicit, genetic differences among individuals contribute to their exposure to drugs and to drug use resulting from drug exposure. Another report from this study found that among twin pairs who volunteered for service in Vietnam, exposure to combat experiences showed significant genetic influence (Lyons et al., 1993).

This study is one of dozens during the past decade that have shown genetic influence on measures that are ostensibly measures of the environment. How can a measure of the environment show genetic influence? If one thinks about the environment as independent of the individual, environmental measures obviously cannot show genetic influence. The weather might affect mood, but weather per se is not inherited. However, measures of the environment are generally *not* independent of the individual. For example, individuals select life-styles that lead to differences in exposure to environmental risks (Rose, 1992). For self-report measures of environmental risks and protective factors, genetic factors can accumulate as such perceptions filter through a person's memories, feelings, and personality. The environment as perceived by individuals is not limited to questionnaires—it is what we mean by experience. Individuals actively construct and reconstruct their experiences, and these processes are influenced by genetic differences among individuals.

How can genetic influence on environmental measures be detected? The simple trick is to treat environmental measures as dependent measures in genetic research designs such as twin and adoption studies. If an environmental measure is not influenced by genetic factors, the correlation for identical twins will be no greater than the correlation for fraternal twins, and correlations for genetically related family members will be no greater than correlations for genetically unrelated members of adoptive families. The surprising finding during the past decade is that twin and adoption studies consistently point to genetic influence for diverse measures of the environment. For example, significant genetic influence has been reported for parent–child interactions, sibling interactions, television viewing, childhood accidents, classroom environments, peer groups, life events, social support, work environments, divorce, education, and socioeconomic status (Plomin, 1994b). Genetic influence on environmental measures is just as real as genetic influence on outcome measures in terms of DNA. That is, as it becomes possible to identify genes associated with complex traits, genes can be identified that are associated with measures of the environment.

Most quantitative genetic research on environmental measures has focused on the family environment, following the lead of the first studies in this area (Rowe, 1981, 1983). For example, an ongoing study specifically designed to investigate genetic influence on a broad range of measures of family environment has reported ubiquitous genetic influence on adolescent children's as well as their parents' perceptions of family environment (Plomin, Reiss, Hetherington, & Howe, 1994) and on ratings of parent–child interactions made from videotape observations (O'Connor, Hetherington, Reiss, & Plomin, 1995). Although family environment is certainly relevant to behavioral medicine, life events and social support may be of greater immediate interest as examples of environmental risks and protective factors.

Life Events

Measures of life events have been used in more than a thousand studies as measures of environmental risk (Holmes, 1979). Typical life events items assess marital or work difficulties, financial problems, illnesses and injuries, and other crises such as being robbed or assaulted. The Swedish Adoption/Twin Study of Aging (SATSA) (Pedersen et al., 1991) included the Social Readjustment Rating Scale (Holmes & Rahe, 1967) modified for older individuals (Persson, 1980). Using the powerful SATSA design of identical and fraternal twins adopted apart (174 pairs) and matched twins reared together (225 pairs) 50 years old and older, the model-fitting estimate of heritability for the total life events score was 40% (Plomin, Lichtenstein, Pedersen, McClearn, & Nesselroade, 1990).

Finding genetic effects on life events implies that such events do not just happen capriciously to individuals. Negative life events happen (or are perceived to happen) to some people more than others. To some extent, this "bad luck" must be related to genetically influenced characteristics of individuals. If this reasoning is correct, it should follow that those events that the individual is involved in and has some control over should show greater heritability than events that are out of the individual's hands. Examples of events that to some extent involve the person include serious conflicts with one's child, major deterioration in financial status, divorce, and paying a fine for a minor violation of the law. Events in which the person is less involved include events that happen to others, such as serious illness in a child, forced change in residence, mental illness of spouse, and death of friends.

SATSA results confirm this expectation. Events that were judged to involve the respondent to a greater degree show greater heritability than events that happen to others. Heritability estimates were 43% for events to self and 18% for events to others. It might seem odd that events that occur to others, such as serious illness of one's child, show any genetic contribution at all. However, it is likely that events such as familial illness entail some genetic effects. Another possibility lies in the use of perceptions of life events. Genetic factors might dispose some dour individuals to consider their child's illness as a "serious illness," in contrast to other people who would not label the same illness as serious.

The life events literature has also made a distinction between the effects of positive and those of negative life events. A keystone of the original Holmes and

Rahe life events instrument was the assumption that stress is imposed by positive events such as marriage as well as by negative events such as divorce. Subsequent research has indicated that negative events are more predictive of problems than are positive events (Thoits, 1983). Nonetheless, SATSA analyses yielded similar results for positive and negative events. Heritability estimates were 31% for positive events and 36% for negative events.

In SATSA, analyses of longitudinal follow-up testing 3 and 6 years after the original wave of testing confirm these basic findings, as do the results of three other twin studies. The first study was a small twin study [41 monozygotic (MZ) pairs and 29 dizygotic (DZ) pairs] that was phrased in terms of affect and depression (Wierzbicki, 1989). Nonetheless, the study included the Life Experiences Survey (Sarason, Johnson, & Siegel, 1978) and the Pleasant and Unpleasant Events Schedules (Lewinsohn & Amenson, 1978). Total frequency of life events on these measures showed significant genetic influence despite the small sample size. The emotional impact of the life events also showed significant genetic influence.

Another study involved reared-apart adult twins [Moster (1990), cited in McGue, Bouchard, Lykken, & Finkel, 1991)]. Significant genetic influence was found, and personal events were substantially more heritable (51%) than events to others (18%). Another replication study was a large classic twin study (2315 pairs) comparing identical and fraternal twins 17–55 years of age who were reared together (Kendler, Neale, Kessler, Heath, & Eaves, 1993). The study included 44 items related to life events during the past year. Significant heritability (26%) was estimated for a total life events score. The results also confirm the SATSA finding that personal events show greater heritability than events to others. For example, the greatest heritability was found for the respondents' own financial problems, whereas death, illness/injury, and crises of others showed nonsignificant heritability. It is noteworthy that the results were generally similar for males and females, although marital difficulties appeared to show genetic effects for males, but nor for females. A twin study of divorce involving 1516 pairs of twins suggested substantial heritability (McGue & Lykken, 1992). When one twin has been divorced, the risk of divorce for the co-twin is 45% for MZ twins and 30% for DZ twins. A preliminary report of another twin study of divorce suggests less heritability, although the sample showed unusually low rates of divorce (Turkheimer, Lovett, Robinette, & Gottesman, 1992).

One twin study of 68 MZ and 109 DZ pairs did not replicate these findings for life events, but this study was different from the others in two respects (McGuffin & Katz, 1993). A checklist of 12 common categories of events during the preceding 6 months was employed. A second difference is that the sample was selected for diagnoses of depression. Perhaps for this reason, the reported frequency of life events was very high, about 70%, for both the depressed probands and their co-twins. No apparent concordance emerged for either MZ or DZ twins for a dichotomous measure of whether any life events were reported, which seems odd given that the frequency of life events was so high for both probands and co-twins. When the number of events was used as a dimension rather than a dichotomy, familial resemblance was observed, but correlations were similar for MZ and DZ twins. The results were not analyzed separately for events to self and events to others.

Although there is no agreement concerning the best way to assess life events, there is widespread dissatisfaction with traditional questionnaire measures (e.g., Paykel, 1983). Finding genetic effects on questionnaire measures of life events justifies the need for research using more expensive instruments such as interviews. One family study of neurotic depression (McGuffin, Katz, & Bebbington, 1988) included an interview measure of life events, the Life Events and Difficulties Schedule (Brown & Harris, 1978). Familial resemblance was found for life events, although it remains to be seen whether familial resemblance for this interview measure is due to genetic factors.

Social Support

Social support also shows genetic influence. Social support refers to the intensiveness and extensiveness of the network of social relationships that surround a person as well as the perceived adequacy of the network (Berkman, 1983). It has become a defining characteristic of "successful aging" (Rowe & Kahn, 1987). In SATSA, a modified version of the Interview Schedule for Social Interaction (Henderson, Duncan-Jones, Byrne, & Scott, 1980) was employed that yielded two scales: quality (perceived adequacy) and quantity or frequency of interactions with relatives and friends (Bergeman, Plomin, Pedersen, McClearn, & Nesselroade, 1990). Significant heritability was found for the quality scale, but not for the quantity scale. A study of adult female twins replicated these findings (Kessler, Kendler, Heath, Neale, & Eaves, 1992).

Consideration of genetic contributions to measures of environmental risks and protective factors suggests new ways of thinking about environmental influences. For example, it provides an empirical basis for exploring the active role of individuals in creating and interpreting their experience (Scarr, 1992). Another example of far-reaching significance concerns the correlates of environmental risk. If environmental risks and protective factors show genetic influence as do measures of biobehavioral health, then it is likely that genetic factors contribute to correlations between them.

CAUSES AND CONSEQUENCES OF GENETIC INFLUENCE ON ENVIRONMENTAL MEASURES

Two programmatic directions for research include the investigation of the causes and the consequences of finding that genetic factors contribute to environmental measures (Plomin & Neiderhiser, 1992). The first issue involves the processes by which genetic factors contribute to environmental measures. For example, to what extent are genetic influences on traditional traits such as personality and psychopathology responsible for the genetic contribution to environmental measures?

The second direction for research considers the most important implication of these findings: Genetic factors might contribute to correlations between environmental measures and behavior. In practice, it is difficult to separate these two issues of causes and consequences because a behavioral correlate of an environmental measure can usually be conceptualized as either an antecedent or a consequence of the environmental measure. For example, is depression the cause or consequence of life events, or both? For this reason, this section reviews relevant research using the agnostic term *correlates* rather than causes or consequences.

Genotype–Environment Correlation

The analyses discussed in this section can be viewed as attempts to identify genotype–environment correlation for specific measures of environment. The term "genotype–environment correlation" literally refers to the correlation between genetic and environmental influences for a particular trait. Three types of genotype–environment correlation have been described: passive, reactive, and active (Plomin, DeFries, & Loehlin, 1977). Passive genotype–environment correlation occurs when children passively inherit from their parents environments that are correlated with their genetic propensities. Reactive genotype–environment correlation refers to experiences of individuals that derive from reactions of other people to their genetic propensities. Active genotype–environment correlation occurs when individuals select, modify, construct, or reconstruct experiences that are correlated with their genetic propensities. A developmental theory of genetics and experience predicts that the reactive and active forms of genotype–environment correlation become more important as children experience environments outside the family and begin to play a more active role in the selection and construction of their experiences (Scarr & McCartney, 1983).

Three methods are available to investigate the contribution of genetic factors to correlations between environment and behavior. These methods differ in the type of genotype–environment correlation that they detect (Plomin, 1994). The first method, comparing correlations between measures of environment and behavior in nonadoptive and adoptive families, focuses on passive genotype–environment correlation. The second method considers reactive and active genotype–environment correlation by correlating experiences of adopted children with characteristics of their biological parents. The third method, multivariate genetic analysis of correlations between environment and behavior, encompasses all three types of genotype–environment correlation. Multivariate genetic analysis is the focus of the following discussion.

Multivariate Genetic Analysis of Genotype–Environment Correlation

As mentioned earlier, multivariate genetic analysis assesses genetic and environmental contributions to the covariance between two measures rather than to the variance of each measure considered separately (Martin & Eaves, 1977). The power of multivariate genetic analysis can be harnessed to study the etiology of associations between environmental measures and their correlates. The essence of estimating genetic and environmental contributions to phenotypic covariance is the cross-correlation. For example, a cross-twin correlation is the correlation between one twin's score on one measure and the other twin's score on another measure. In the present case, the issue is the correlation between one twin's score on an environmental measure and the other twin's score on a behavioral measure. Everything else is similar to the usual univariate analysis of the variance of a single measure. In univariate twin analysis, doubling the difference between identical and fraternal twin correlations provides a rough estimate of heritability, the proportion of phenotypic variance attributed to genetic variance. In bivariate analysis, doubling the difference between identical and fraternal cross-twin correlations estimates the genetic contribution to the phenotypic correlation. In practice, multivariate genetic analysis is conducted using model-fitting approaches (Neale & Cardon, 1992).

Only a few multivariate genetic analyses between measures of the environment and behavior have been reported (Plomin, 1994b), but these analyses suggest that such associations are mediated genetically to some extent. For example, measures of life events correlate modestly with depression. To what extent is this correlation due to genetic overlap between life events and depression? A multivariate genetic analysis between life events and depression in SATSA suggested that genetic factors contribute to the correlation (Neiderhiser, Plomin, Lichtenstein, Pedersen, & McClearn, 1992). Social support also correlates with depression and life satisfaction, and again SATSA analyses indicate that genetic factors common to social support and to the behavioral measures contribute to their correlation (Bergeman, Plomin, Pedersen, & McClearn, 1991), a finding confirmed in another twin study (Kessler et al., 1992). A particularly surprising finding is that socioeconomic status and self-reported health are both heritable and correlate in part for genetic reasons (Lichtenstein, Harris, Pedersen, & McClearn, 1995).

The main limitation in research on mediation of genetic influence on environmental measures has been trying to find substantial correlations between environmental measures and behavioral measures (Neiderhiser, 1994). For example, multivariate genetic analyses between life events and personality were uninformative because the phenotypic correlation between them is so low (Plomin et al., 1990). This presents a special problem for the analysis of genetic mediation of environment–behavior associations. If the magnitude of the association between environment and outcome is low, there is not much covariance to partition into genetic and environmental components. Nonetheless, when correlations of reasonable magnitude have been found between measures of environment and behavior, genetic factors appear to contribute to the correlations.

CONCLUSIONS AND IMPLICATIONS

One of the several important advances in genetic research that go beyond heritability is the analysis of genetic influence on environmental measures and on associations between environmental measures and behavior. Genetic factors have been shown to contribute to measures of the environment relevant to research in behavioral medicine. The most important implication of finding genetic contributions to measures of the environment is that genetic factors might contribute to the prediction of behavior from such environmental measures. The first studies in this area support the hypothesis, as in the example of genetic mediation of the association between social support and depression.

In considering the implications of these findings, two caveats should be mentioned. First, this research does not mean that the environment is not environmental in its effects on behavior. The research clearly indicates that most of the variance in environmental measures is indeed environmental in origin. Second, these findings do not imply any necessary limitations on our ability to intervene successfully in the public health sector or in the clinician's office. Heritability does not imply immutability or untreatability. This is an important point, because finding genetic influence on environmental measures has provoked outrage among psychologists who fear it will undermine attempts to improve life circumstances (e.g., Baumrind, 1993). However, genetic influence on anything, including environmental risk factors, does not imply immutability. As Rutter, Champion, Quinton, Maughan, and Pickles (1993) indicates:

> If we are to intervene effectively, it is crucially important that we understand how environmental risks arise. . . . If it turns out that genetic factors play an important part in that causal chain, so be it. However, even if that were to prove to be the case, that does not necessarily mean that interruption of the chain is not open to environmental manipulation. Much depends on the ways in which the genetic factors operate. . . . As medical geneticists have pointed out recently . . . much of the importance of modern genetic research, using molecular genetic techniques, is likely to lie in the identification and understanding of environmental risk mechanisms.

Although this research is too new to have immediate implications for the clinician confronted with a specific individual's problems, it puts a new slant on some old issues of prevention and intervention. At the simplest level, if genetic differences among individuals contribute to their exposure to environmental risk, attempts to change environmental risks in highly controlled experimental situations may be thwarted in the real world, where people are free to choose their own poison. It might be useful to consider the Dawkins (1983) concept of the "extended phenotype." Genetic factors contribute not only to a disorder "in the individual" but also to the individual's active construction of experience correlated with genetic propensities. That is, some individuals are at risk for certain experiences that in turn put them at risk for disorders. In terms of prevention, it might be useful to consider not just environmental risk factors and genetic risk factors, but also genotype–environment risk factors that lie in the correlation between nature and nurture. It might also be helpful to think about intervention in terms of changing genotype–environment correlations rather than changing genetic dispositions or environmental risks.

These findings are more immediately relevant for basic theory and research. The most fundamental implication is that environmental measures cannot be assumed to be environmental just because they are called environmental. Indeed, research to date suggests that it is safer to assume that ostensible measures of the environment include genetic effects. Taking this argument further, a recent book has concluded that environmental research has been fundamentally flawed because it has failed to take into account genetic confounds (Rowe, 1994).

Environmental theory has moved away from passive models in favor of models that recognize the active role of individuals in selecting, modifying, and creating their own environments (Wachs, 1992). Despite this shift in environmental theory, research seldom addresses this active model of environments. The genetic contribution to individual differences in experience provides an opportunity to explore these processes empirically.

Three concrete examples involve measurement of the environment. First, extant environmental measures remain much more passive than active despite the shift from passive to active models of environmental influence. Progress in this field depends on developing measures of the environment that reflect the active role of the individual in constructing and reconstructing experience. For example, it seems likely that environmental measures that assess individuals' active engagement with their environments will show even greater evidence for genetic influence.

Second, it might be possible to develop measures of the environment that show less genetic influence. Although genetic influence does not imply immutability, it seems reasonable to suggest that environmental processes less influenced by genetic differences among individuals will be more amenable to prevention and intervention.

Third, genetic research highlights the need to identify strong environment–behavior associations, not merely statistically significant associations. Genetic analysis of environment–behavior associations has been hampered by the difficulty of finding associations of sufficiently large effect size to permit investigation of the genetic contribution to the covariation between environment and outcome.

Finally, these findings have implications for behavioral assessment. If genetic influence affects how individuals select, modify, and even create environmental risks and protective factors, assessing individuals' reactions in a standard situation might miss these processes. This issue has been illustrated in relation to cardiovascular reactivity (Turner, 1994, p. 126):

> We are largely concerned with the individual differences—for example, in blood pressure response—that a standard situation elicits from people. In that situation, some individuals will show large responses while others will show minimal change in physiological activation. The implicit assumption is that the high reactor, as compared to the medium or the low reactor, is the one at greatest risk of later cardiovascular disease, since he or she shows the greatest propensity to react. However, this scenario does not pay attention to how likely it is that a particular individual will be exposed to real-life (naturalistic) stressful situations . . . some people actively seek stressful situations, while others do not.

Cardiovascular reactivity displayed during stress and the propensity to seek or create stressful situations might be independent genetically. It seems reasonable to propose that the latter might be more predictive of cardiovascular disease. A related argument has been made in discussing models of associations between personality and disease: Personality might precipitate dangerous behaviors that expose individuals to risk for disease (Suls & Rittenhouse, 1990).

In conclusion, theory and research in genetics (nature) and in environment (nurture) are beginning to converge. Although there is still a very long way to go, the common ground is a model of active organism–environment interaction in which nature and nurture play a duet rather than one directing the performance of the other. It seems clear that some of the most interesting questions for genetic research involve the environment and some of the most interesting questions for environmental research involve genetics. It is time to put the nature–nurture controversy behind us and to bring nature and nurture together in order to understand the processes by which genotypes become phenotypes.

ACKNOWLEDGMENTS. Preparation of this chapter and some of the research it reviews was supported by grants from the National Institute of Aging (AG04563), the National Institute of Child Health and Human Development (HD10333, HD18426), the National Institute of Mental Health (MH43899, MH43373), and the National Science Foundation (SBR-9108744).

REFERENCES

Baumrind, D. (1993). The average expectable environment is not good enough: A response to Scarr. *Child Development, 64,* 1299–1317.

Bergeman, C. S., Plomin, R., Pedersen, N. L., & McClearn, G. E. (1991). Genetic mediation of the relationship between social support and psychological well-being. *Psychology and Aging, 6,* 640–646.

Bergeman, C. S., Plomin, R., Pedersen, N. L., McClearn, G. E., & Nesselroade, J. R. (1990). Genetic and environmental influences on social support: The Swedish Adoption/Twin Study of Aging (SATSA). *Journal of Gerontology, 45,* 101–106.

Berkman, L. F. (1983). The assessment of social networks and social support in the elderly. *Journal of the American Geriatrics Society, 31,* 743–749.

Brown, G. W., & Harris, T. (1978). *The social origins of depression.* London: Tavistock.

Cambien, F., Poirier, O., Lecerf, L., Evans, A., Cambou, J. P., Arveiler, D., Luc, G., Bard, J. M., Bara, L., Ricard, S., Tiret, L., Amouyel, P., Alhenc-Gelas, F., & Soubrier, F. (1992). Deletion polymorphism in the gene coding for angiotensin-converting enzyme is a potent risk factor for myocardial infarction. *Nature, 359,* 641–644.

Dawkins, R. (1983). *The extended phenotype: The long reach of the gene*. Oxford: Oxford University Press.

DeFries, J. C., & Fulker, D. W. (1985). Multiple regression analysis of twin data. *Behavior Genetics, 16,* 1–10.

Dunn, J., & Plomin, R. (1990). *Separate lives: Why siblings are so different*. New York: Basic Books.

Eaves, L. J., Eysenck, H. J., & Martin, N. G. (1989). *Genes, culture and personality: An empirical approach*. New York: Academic Press.

Eaves, L. J., Long, J., & Heath, A. C. (1986). A theory of developmental change in quantitative phenotypes applied to cognitive development. *Behavior Genetics, 16,* 143–162.

Edwards, J. H. (1991). The formal problems of linkage. In P. McGuffin & R. Murray (Eds.), *The new genetics of mental illness* (pp. 58–70). London: Butterworth-Heinemann.

Friedman, H. S. (Ed.) (1990). *Personality and disease*. New York: Wiley Interscience.

Fulker, D. W., & Cardon, L. (1994). A sib-pair approach to interval mapping of quantitative trait loci. *American Journal of Human Genetics, 54,* 1092–1097.

Gelderman, H. (1975). Investigations on inheritance of quantitative characters in animals by gene markers. I. Methods. *Theoretical and Applied Genetics, 46,* 319–330.

Grilo, C. M., & Pogue-Geile, M. F. (1991). The nature of environmental influences on weight and obesity: A behavior genetic analysis. *Psychological Bulletin, 110,* 530–537.

Henderson, D., Duncan-Jones, P., Byrne, D. G., & Scott, R. (1980). Measuring social relationships: The Interview Schedule for Social Interaction. *Psychological Medicine, 10,* 723–734.

Hewitt, J. K. (1993). The new quantitative genetic epidemiology of behavior. In R. Plomin & G. E. McClearn (Eds.), *Nature, nurture, and psychology* (pp. 401–415). Washington, DC: American Psychological Association.

Hewitt, J. K., Eaves, L. J., Neale, M. C., & Meyer, J. (1988). Resolving causes of developmental continuity or "tracking": Longitudinal studies during growth. *Behavior Genetics, 18,* 133–151.

Hobbs, H. H., Russell, D. W., Brown, M. S., & Goldstein, J. L. (1990). The LDL receptor locus in familial hypercholesterolemia: Mutational analysis of a membrane protein. *Annual Review of Genetics, 24,* 133–170.

Holmes, T. H. (1979). Development and application of a quantitative measure of life change magnitude. In J. E. Barrett (Ed.), *Stress and mental disorder* (pp. 37–53). New York: Raven Press.

Holmes, T. H., & Rahe, R. H. (1967). The Social Readjustment Rating Scale. *Journal of Psychosomatic Research, 11,* 213–218.

Humphries, S. E. (1988). DNA polymorphisms of the apolipoprotein genes: Their use in the investigation of the genetic component of the hyperlipidaemia and atherosclerosis. *Atherosclerosis, 72,* 89–108.

Kendler, K. S., Neale, M., Kessler, R., Heath, A., & Eaves, L. (1993). A twin study of recent life events and difficulties. *Archives of General Psychiatry, 50,* 789–796.

Kessler, R. C., Kendler, K. S., Heath, A., Neale, M. C., & Eaves, L. J. (1992). Social support, depressed mood, and adjustment to stress: A genetic epidemiologic investigation. *Journal of Personality and Social Psychology, 62,* 257–272.

King, R. A., Rotter, J. I., & Motulsky, A. G. (Eds.) (1992). *The genetic basis of common diseases*. New York: Oxford University Press.

Lewinsohn, P. M., & Amenson, C. S. (1978). Some relations between pleasant and unpleasant mood-related events and depression. *Journal of Abnormal Psychology, 87,* 644–654.

Lichtenstein, P., Harris, J. R., Pedersen, N. L., & McClearn, G. E. (1995). Socioeconomic status and physical health, how are they related? An empirical study based on twins reared apart and twins reared together. *Social Science and Medicine* (in press).

Lyons, M. J., Goldberg, J., Eisen, S. A., True, W., Tsuang, M. T., Meyer, J. M., & Henderson, W. G. (1993). Do genes influence exposure to trauma? A twin study of combat. *American Journal of Medical Genetics (Neuropsychiatric Genetics), 48*, 22–27.

Martin, N. G., & Eaves, L. J. (1977). The genetical analysis of covariance structure. *Heredity, 38*, 79–95.

McGue, M., Bouchard, T. J., Lykken, D. T., & Finkel, D. (1991). On genes, environment, and experience. *Behavioral and Brain Sciences, 14*, 400–401.

McGue, M., & Lykken, D. T. (1992). Genetic influence on risk of divorce. *Psychological Science, 3*, 368–373.

McGuffin, P., & Katz, R. (1993). Genes, adversity and depression. In R. Plomin & G. E. McClearn (Eds.), *Nature, nurture, and psychology* (pp. 217–230). Washington, DC: American Psychological Association.

McGuffin, P., Katz, R., & Bebbington, P. (1988). The Camberwell Collaborative Depression Study. III. Depression and adversity in the relatives of depressed probands. *British Journal of Psychiatry, 152*, 775–782.

Morrison, N. A., Qi, J. C., Tokita, A., Kelly, P. J., Crofts, L., Nguyen, T. V., Sambrook, P. N., & Eisman, J. A. (1994). Prediction of bone density from vitamin D receptor alleles. *Nature, 367*, 284–287.

Moster, M. (1990). Stressful life events: Genetic and environmental components and their relationship to affective symptomatology. Unpublished doctoral dissertation. Minneapolis: University of Minnesota.

Motulsky, A. G., & Brunzell, J. D. (1992). The genetics of coronary atherosclerosis. In R. A. King, J. I. Rotter, & A. G. Motulsky (Eds.), *The genetic basis of common diseases* (pp. 150–169). New York: Oxford University Press.

Neale, M. C., & Cardon, L. R. (1992). *Methodology for genetic studies of twins and families*. Norwell, MA: Kluwer Academic Publishers.

Neiderhiser, J. M. (1994). Family environment in early childhood and outcomes in middle childhood: Genetic mediation. In J. C. DeFries, R. Plomin, & D. W. Fulker (Eds.), *Nature and nurture during middle childhood* (pp. 249–261). Cambridge, MA: Blackwell.

Neiderhiser, J. M., Plomin, R., Lichtenstein, P., Pedersen, N. L., & McClearn, G. E. (1992). The influence of life events on depressive symptoms over time. *Behavior Genetics, 22*, 740 (abstract).

O'Connor, T. G., Hetherington, E. M., Reiss, D., & Plomin, R. (1995). A twin-family study of observed parent–adolescent interactions. *Child Development* (in press).

Owen, M. J., & McGuffin, P. (1993). Association and linkage: Complementary strategies for complex disorders. *Journal of Medical Genetics, 30*, 638–639.

Paykel, E. S. (1983). Methodological aspects of life events research. *Journal of Psychosomatic Research, 27*, 341–352.

Pedersen, N. L., McClearn, G. E., Plomin, R., Nesselroade, J. R., Berg, S., & DeFaire, U. (1991). The Swedish Adoption/Twin Study of Aging: An update. *Acta Geneticae Medicae et Gemellologiae, 40*, 7–20.

Persson, G. (1980). Prevalence of mental disorders in a 70-year-old urban population. *Acta Psychiatrica Scandinavica, 62*, 112–118.

Plomin, R. (1986). *Development, genetics, and psychology*. Hillsdale, NJ: Erlbaum Associates.

Plomin, R. (1994a). Genetic research and identification of environmental influences: Emanuel Miller Memorial Lecture (1993). *Journal of Child Psychology and Psychiatry, 36*, 33–68.

Plomin, R. (1994b). *Genetics and experience: The interplay between nature and nurture*. Newbury Park, CA: Sage Publications.

Plomin, R., Chipuer, H. M., & Neiderhiser, J. M. (1994). Behavioral genetic evidence for the importance of nonshared environment. In E. M. Hetherington, D. Reiss, & R. Plomin (Eds.), *Separate social worlds of siblings: The impact of nonshared environment on development* (pp. 1–31). Hillsdale, NJ: Erlbaum Associates.

Plomin, R., & Daniels, D. (1987). Why are children in the same family so different from one another? *Behavioral and Brain Sciences, 10,* 1–16.

Plomin, R., DeFries, J. C., & Loehlin, J. C. (1977). Genotype–environment interaction and correlation in the analysis of human behavior. *Psychological Bulletin, 84,* 309–322.

Plomin, R., Lichtenstein, P., Pedersen, N. L., McClearn, G. E., & Nesselroade, J. R. (1990). Genetic influence on life events during the last half of the life span. *Psychology and Aging, 5,* 25-30.

Plomin, R., & McClearn, G. E. (Eds.) (1993). *Nature, nurture, and psychology.* Washington, DC: American Psychological Association.

Plomin, R., & Neiderhiser, J. M. (1992). Genetics and experience. *Current Directions in Psychological Science, 1,* 160–163.

Plomin, R., Owen, M. J., & McGuffin, P. (1994). The genetic basis of complex human behaviors. *Science, 264,* 1733–1739.

Plomin, R., Reiss, D., Hetherington, E. M., & Howe, G. (1994). Nature and nurture: Genetic influence on measures of the family environment. *Developmental Psychology, 30,* 32–43.

Plomin, R., & Rende, R. (1991). Human behavioral genetics. *Annual Review of Psychology, 42,* 161–190.

Risch, N. (1990). Linkage strategies for genetically complex traits. II. The power of affected relative pairs. *American Journal of Human Genetics, 46,* 229–241.

Rose, R. J. (1992). Genes, stress, and cardiovascular reactivity. In J. R. Turner, A. Sherwood, & K. C. Light (Eds.), *Individual differences in cardiovascular response to stress* (pp. 87–102). New York: Plenum Press.

Rowe, D. C. (1981). Enviornmental and genetic influences on dimensions of perceived parenting: A twin study. *Developmental Psychology, 17,* 203–208.

Rowe, D. C. (1983). A biometrical analysis of perceptions of family environment: A study of twins and singleton sibling kinships. *Child Development, 54,* 416–423.

Rowe, D. C. (1994). *The limits of family influence.* New York: Guilford Press.

Rowe, J. W., & Kahn, R. L. (1987). Human aging: Usual and successful. *Science, 237,* 143–149.

Rutter, M., Champion, L., Quinton, D., Maughan, B., & Pickles, A. (1995). Origins of individual differences in environmental risk exposure. In P. Moen, G. Elder, & K. Luscher (Eds.), *Perspectives on the ecology of human development.* Ithaca, NY: Cornell University Press.

Sarason, I. G., Johnson, J. H., & Siegel, J. M. (1978). Assessing the impact of life changes: Development of the Life Experiences Survey. *Journal of Consulting and Clinical Psychology, 46,* 932–946.

Scarr, S. (1992). Developmental theories for the 1990s: Development and individual differences. *Child Development, 63,* 1–19.

Scarr, S., & McCartney, K. (1983). How people make their own environments: A theory of genotype → environment effects. *Child Development, 54,* 424–435.

Sing, C. F., & Boerwinkle, E. A. (1987). Genetic architecture of inter-individual variability in apolipoprotein, lipoprotein and lipid phenotypes. In G. Bock & G. M. Collins (Eds.), *Molecular approaches to human polygenic disease* (pp. 99–122). Chichester, England: John Wiley.

Sobell, J. L., Heston, L. L., & Sommer, S. S. (1992). Delineation of the genetic predisposition to a multifactorial disease; A general approach on the threshold of feasibility. *Genomics, 12,* 1–6.

Stunkard, J., Foch, T. T., & Hurbec, Z. (1986). A twin study of human obesity. *Journal of the American Medical Association, 256,* 51–54.

Suls, J., & Rittenhouse, J. D. (1990). Models of linkages between personality and disease. In H. S. Friedman (Ed.), *Personality and disease* (pp. 38–64). New York: John Wiley.

Thoits, P. A. (1983). Dimensions of life events that influence psychological distress: An evaluation and synthesis of the literature. In H. A. Kaplan (Ed.), *Psychological stress: Trends in theory and research* (pp. 33–103). New York: Academic Press.

Tsuang, M. T., Lyons, M. J., Eisen, S. A., True, W. T., Goldberg, J., & Henderson, W. (1992). A twin study of drug exposure and initiation of use. *Behavior Genetics, 22,* 756 (abstract).

Turkheimer, E., Lovett, G., Robinette, C. D., & Gottesman, I. I. (1992). The heritability of divorce: New data and theoretical implications. *Behavior Genetics, 22,* 757 (abstract).

Turner, J. R. (1994). *Cardiovascular reactivity and stress: Patterns of physiological response.* New York: Plenum Press.

Wachs, T. D. (1992). *The nature of nurture.* Newbury Park, CA: Sage Publications.

Wierzbicki, M. (1989). Twins' responses to pleasant, unpleasant, and life events. *Journal of Genetic Psychology, 150,* 135–145.

CHAPTER THIRTEEN

Quantitative Trait Loci

Mapping Genes for Complex Traits

LON R. CARDON

INTRODUCTION

The preceding chapters demonstrate the utility of behavior genetic methodology and the twin/family study design for determining etiological components of traits of medical and behavioral importance. Knowledge of the relative influence of genetic and environmental determinants, their developmental patterns, their different manifestations in males and females, and their commonality across traits is essential for accurate characterization of most traits of interest and for development of effective strategies for screening and intervention.

From an environmental perspective, the study of twins or their families or both provides an exceptionally well-controlled design, a paradoxical asset of twins given their usual association with genetic hypotheses. By controlling for heritable variation that contributes to individual differences, the investigator has less total variation to explain and therefore may evaluate salient features of the environment with more precision. The advantages from a genetic viewpoint of analyzing informative family relationships such as twins are elucidated by example throughout this text. It is clear that genetic analyses of twins have moved beyond simple, but necessary, preliminary questions such as "Is blood pressure heritable?" or "Is obesity genetic?" to more searching questions such as "Are the genes that cause variability in cardiovascular responses to stress the same in males and females?" and "Do the same genes contribute to obesity all through

---

LON R. CARDON • Sequana Therapeutics, 11099 North Torrey Pines Road, Suite 160, La Jolla, California 92037.

Behavior Genetic Approaches in Behavioral Medicine, edited by J. Rick Turner, Lon R. Cardon, and John K. Hewitt. Plenum Press, New York, 1995.

life, or are new genes expressed at some point in development?" These types of questions, empirically tested, obviously have much to contribute to our understanding of individual differences.

In the genetic framework, however, these questions are not the end points of our interest. The ultimate goal is to locate the genes that influence complex diseases and behavior, define the differences that are ultimately expressed in the phenotypes, and understand the pathways that lead to phenotypic expression. Unfortunately, achieving all aspects of this goal is far beyond the reach of current methodology and epidemiological understanding for many characters. However, the first step toward this ultimate objective, identifying the genes, is receiving considerable attention and has yielded some very promising results. Because of intense interest in mapping genes for complex traits, the body of work in this broad area is not easily or fruitfully summarized as a whole. In this chapter, we provide a brief overview of the issues involved in such studies, some of the methods employed, and some illustrative examples whereby initial research into genetic and environmental etiologies using twins and families has contributed to the subsequent success of mapping the genetic loci.

GENOTYPE MEASUREMENT AND LINKAGE

Mapping strategies move a step beyond the inferential twin design with respect to information employed in analysis. Whereas twin studies infer the effects of a combination of genetic factors, typically modeled as latent factors in the structural equation approach (Neale & Cardon, 1992), studies aimed at locating genes incorporate direct measurements of specific chromosomal regions. There are very many known loci throughout the human chromosomes that can be directly observed using molecular genetic procedures (NIH/CEPH, 1992). Some of these comprise known genes, but many are simply sequences of deoxyribonucleic acid (DNA) without any known function. Such loci are known as chromosomal "markers" that serve to delineate physical regions in the human genome (all 23 chromosomes). The most useful of these markers for genetic studies contain many alternate forms, known as "polymorphisms," that arise for a number of reasons, including mutations or variability in the number of repeated DNA base pairs. Each different form of a gene marker is termed an "allele." Historical markers include the ABO blood group and serum enzymes; more recent markers are based on advances in molecular genetics that provide more polymorphic systems (Innis, Gelfand, Sninskey, & White, 1990).

Classic genetic strategies test the assumption that a given marker of known location is situated near the unknown gene causing the trait of interest, so that the marker and gene are transmitted together to offspring (Ott, 1991). The extent to which the marker and gene cosegregate is referred to as "linkage." If the marker and gene are not positioned very close to one another on the chromosome, they may be separated during the process of meiosis, in which matched pairs of chromosomes repeatedly cross over each other and exchange segments of DNA.

The event of chromosomal exchange during crossing over is known as "recombination." If crossing over occurs between the marker and the gene locus, recombination will separate the two so that offspring with the disease gene polymorphism will no longer carry the marker polymorphism found in the parent. Evidence for linkage, which is clearly more easily obtainable with close marker–gene relationships than with those more distant, is derived using statistical procedures that take into account marker–gene cosegregation as a function of recombination and the mode of transmission of the trait.

Classic methods for linkage analysis make use of family data from extended pedigrees, in which the phenotypes and markers are measured in individuals from multiple generations. In this way, one can trace the cosegregation of the trait and marker along familial lineages. Most of the recent successes in linkage analysis have employed this strategy [e.g., susceptibility genes for colon cancer (Bodmer et al., 1987) and melanoma (Cannon-Albright et al., 1992)]. A complete treatment of linkage methods may be found in Ott (1991).

COMPLEX PHENOTYPES

One difficulty in locating genes for behaviors and many diseases relates to the phenotypes themselves. Most of the characters we are interested in examining in twins and families are continuously distributed, e.g., blood pressure, cholesterol, body fat patterns, substance use, and intelligence. While the variability in these traits is precisely the feature that makes them most interesting in terms of individual differences, it also leads to difficulties in finding the genes that determine the variability. The classic methods of gene identification were initially developed for discrete traits under assumptions of single-gene transmission, where one can make use of information about the mode of transmission (e.g., recessive, dominant) following simple Mendelian principles. However, complex traits, by definition, do not necessarily follow strictly monogenic (single-gene) Mendelian patterns; rather, they are multifactorial in nature, involving a number of additive genes, each with relatively small effect. In classic quantitative genetics literature, such traits are referred to as "polygenic" characters. Today, investigators contrast these types of traits with "oligogenic" characters, which involve a relatively small number of distinguishable genes with larger effect size.

Polygenic or oligogenic traits have the potential for a number of events that can mask the Mendelian transmission of each locus, including interactions between gene loci (epistasis) or the same phenotype resulting from expression of different genes or gene combinations (genetic heterogeneity). Of course, the involvement of multiple genes also magnifies the potential for interactions with environmental factors. Multiple genetic loci that are thought to influence continuous traits are known as "quantitative trait loci" (QTLs) (Gelderman, 1975). In experimental nonhuman animals, QTLs of very small effect size—e.g., with single-locus heritabilities of 1–2%—are reasonable candidates for mapping.

Unfortunately, in humans, detection of such small effects is generally much less likely due to sample size and marker constraints. More appropriate targets at present involve oligogenic traits with locus heritabilities in the range of 10–50% (Carey & Williamson, 1991).

Age-related changes also contribute to the difficulty of mapping genes for complex traits. Schork and Weder (1995) have described a model organism in which a number of genes are expressed at different ages throughout development. For example, if the relevant age of expression is not accounted for by sampling across age strata, localization of any gene is very difficult because the trait displays symptoms of genetic heterogeneity. Similar difficulties may arise for cases of sex-limited genetic effects. These are areas in which inferential genetic studies such as those involving twins can be of considerably utility.

SIB-PAIR METHODS FOR QUANTITATIVE TRAIT LOCUS LINKAGE

Sib-pair strategies are perhaps the most widely used methods for analyzing continuous oligogenic traits (Penrose, 1938; Haseman & Elston, 1972; Hill, 1974). In these designs, trait and marker data are obtained from siblings and (optimally) their parents in a number of different families, rather than from large multigenerational pedigrees with many affected individuals. The methods do not involve any assumptions concerning the mode of transmission and are robust with respect to genetic heterogeneity.

The method of Haseman and Elston (1972) exemplifies the sib-pair approach for continuous traits. The simple idea behind this approach is that under linkage to a QTL, differences between siblings in their phenotypes will decrease in accordance with greater similarity at the marker locus. Haseman and Elston employ the proportion of alleles that siblings share identical-by-descent (ibd) as their measure of marker resemblance, which they call π. The ibd relationship requires information concerning the ancestral origins of each allele (e.g., for a sibling with alleles A and a at a marker, it is known from which parent each allele was transmitted). Haseman and Elston formalized this sib-pair procedure with a simple regression equation:

$$Y = \alpha + \beta\pi \tag{1}$$

where Y is the squared difference between the phenotypes of two siblings. In this expression, β provides an estimate of the genetic variance of the QTL ($\sigma^2 g$) as a function of the recombination fraction (θ):

$$\beta = -2\sigma_g^2 (1 - 2\theta)^2$$

This simple regression formulation is illustrated for simulated observations for 1000 sib-pairs in Fig. 1.

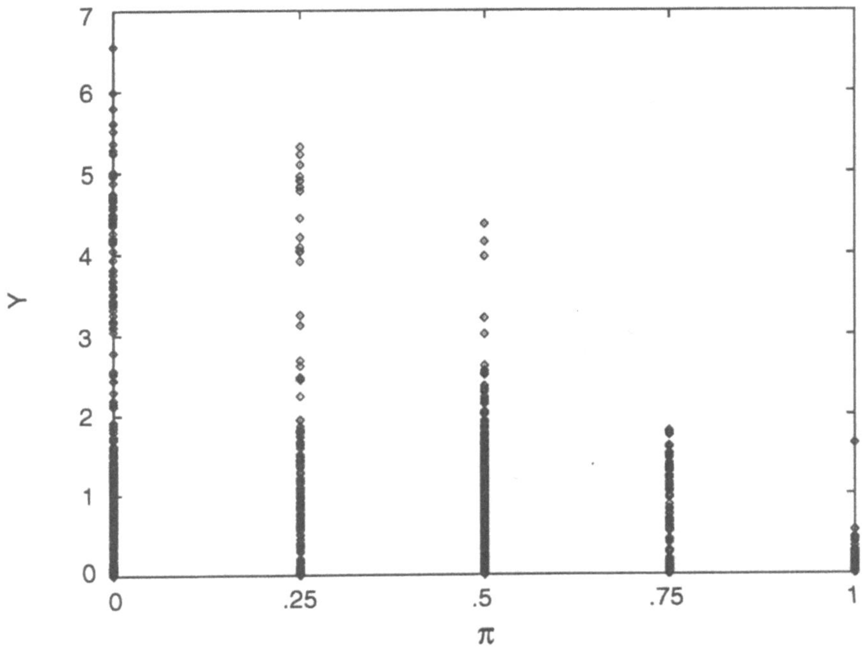

FIGURE 1. Scatterplot of squared differences between sibling phenotypes (Y) as a function of the proportion of alleles shared identical-by-descent at a trait locus (π). The differences between siblings decrease as their number of shared alleles increases, providing evidence for linkage to a QTL.

If the marker and QTL are tightly linked, then $\theta = 0$ and the heritability of the QTL may be estimated as:

$$h^2_{QTL} = -1/2 \frac{\beta}{\sigma^2_p} = \frac{\sigma^2_g}{\sigma^2_p}$$

This heritability is analogous to that obtained from twins, the only difference being that the twin heritability is a function of the total effects of all (unknown) loci. This illustrates the close correspondence between the actual units of inheritance, DNA, as reflected in marker polymorphisms, and the genetic effects that are inferred from twin studies in which no genetic material is obtained. Indeed, the polygenic model that forms the basis of the biometrical twin design is rooted in the effects of individual loci. Algebraic descriptions of the simple relationship between monogenic and polygenic genetic variation are given in Mather and Jinks (1971), Falconer (1990), and Neale and Cardon (1992).

Fulker et al. (1991) developed another sib-pair procedure for QTL linkage that is closely related to the inferential studies of twins and families. The method is based on an approach developed specifically for selected samples of monozygotic (MZ) and dizygotic (DZ) twins (DeFries & Fulker, 1985), in which the

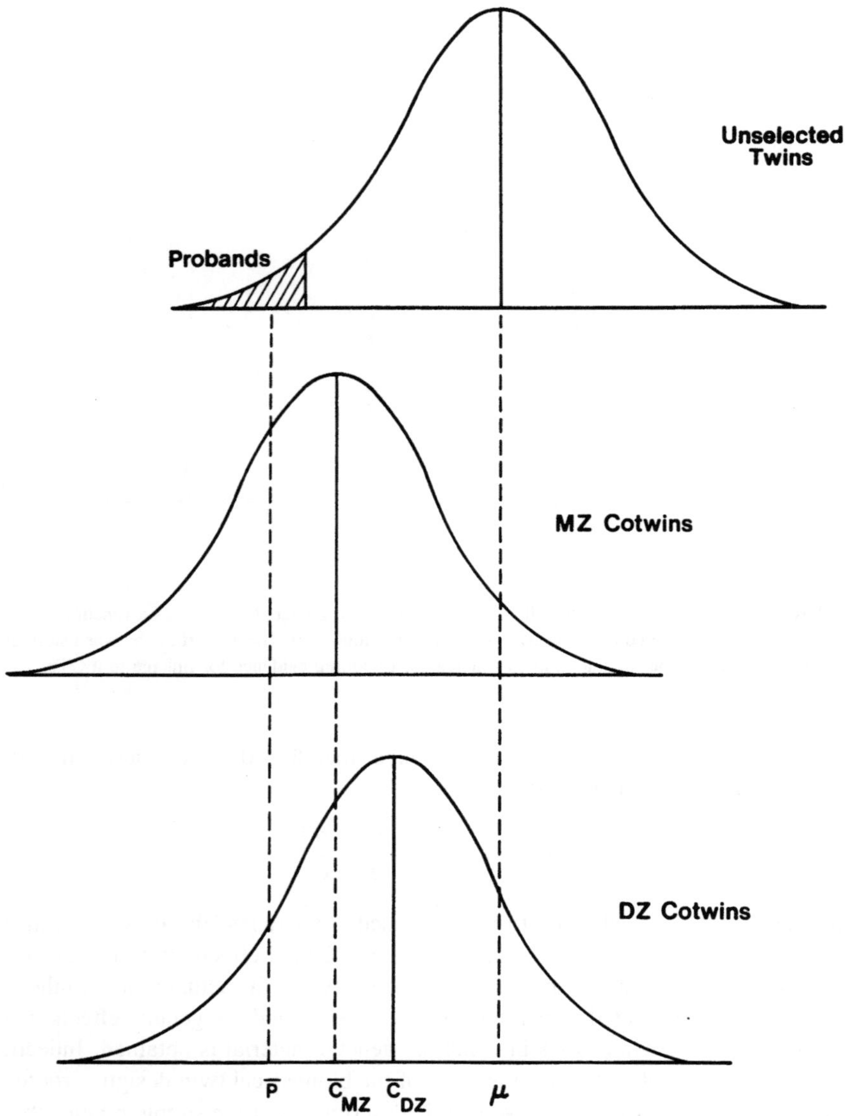

FIGURE 2. Hypothetical distributions of a continuous trait for an unselected sample of twins and for the identical (MZ) and fraternal (DZ) co-twins of probands selected for extreme scores on the trait. The differential regression of the MZ and DZ co-twin means toward the mean of the unselected population (μ) provides a test of genetic etiology. Reprinted by permission from J. C. DeFries, D. W. Fulker, & M. C. LaBuda, "Evidence for a Genetic Aetiology in Reading Disability of Twins," *Nature,* vol. 329, pp. 537–539. Copyright 1987 Macmillian Magazines Ltd.

study design involves ascertainment of individuals having extreme trait scores (e.g., hypertensive twins defined on the basis of blood pressure measurements). The simple idea behind this procedure is that the mean trait scores of co-twins of selected probands should differentially regress back toward the population mean as a function of the heritability of the trait. That is, for DZ twins, who share half their segregating genes on average, co-sibs of selected probands should regress further back to the population mean than should co-sibs of MZ twins, who are genetically identical at all loci.

The principle of this method may be seen in Fig. 2, in which the distribution at the top of the figure represents that of the base population, with a selected group of low-scoring individuals at the tail of the distribution labeled "Probands." The distributions of the MZ and DZ co-twins are shown below and can be seen to have means regressing back toward that of the population, μ. The extent of this regression is greater for DZ than for MZ pairs. The following equation captures this differential regression:

$$C = A + B_1P + B_2R \tag{2}$$

where C is the co-sib score, R is the coefficient of genetic resemblance ($1/2$ for DZ, 1 for MZ), and P is the proband score. DeFries and Fulker (1988) have shown that $B_2 = 2[(\bar{C}_{MZ}-\bar{C}_{DZ})-B_1(\bar{P}_{MZ}-\bar{P}_{DZ})]$, which is equal to $2(\bar{C}_{MZ}-\bar{C}_{DZ})$ when the MZ and DZ probands have the same mean, which is usually expected to be the case. This regression coefficient, when divided by the selection differential, $\bar{P}-\mu$, is an estimate of the heritability of the proband deficit, h^2_g. DeFries and Fulker (1985, 1988) also developed an augmented model for use with unselected or selected samples. This model involves the addition of a product term, PR, representing the product of proband score and the coefficient of relationship:

$$C = A + B_3P + B_4R + B_5PR \tag{3}$$

In this form, the coefficient B_5 is a direct estimate of h^2 and B_3 is a direct estimate of shared environmental variance, c^2.

In addition to providing a very simple method for estimating trait heritability, these procedures easily take into account covariates such as gender and age by the inclusion of appropriate regression terms (Fulker et al., 1991). Recent extensions of this method have involved bivariate regressions, thereby estimating the correlation between genetic factors that influence two traits (Gillis Light and DeFries, 1994).

The extension of these models to sib-pair linkage is straightforward. All that is involved is a substitution of the proportion of alleles shared ibd by two siblings, π, for the coefficient of genetic resemblance, R. Thus, the basic model in equation (2) becomes

$$C = A + B_1P + B_2\pi \tag{4}$$

and the augmented model in equation (3) becomes

$$C = A + B_3P + B_4\pi + B_5P\pi \tag{5}$$

Fulker et al. (1991) and Carey and Williamson (1991) showed that B_2 in equation (4) and B_5 in equation (5) provide direct tests of QTL linkage as a function of the recombination fraction $(1 - 2\theta)^2$. This method provides a useful alternative to the Haseman and Elston (1972) procedure. It is more robust to outlier effects because it uses co-sib and proband values rather than a squared difference quantity, *Y*, which can magnify outlier effects (Cardon & Fulker, 1994). Extensions have been developed to take into account information from multiple markers simultaneously, which strengthens the statistical power of the method (Fulker & Cardon, 1994; Cardon & Fulker, 1994; Fulker, Cherny, & Cardon, 1995; Schork, Amos, Fulker, & Cardon, 1995). These approaches have proven quite successful in mapping genes for reading disability (Cardon et al., 1995).

It should be noted that the sib-pair method, while flexible and robust, suffers from a lack of statistical power to detect QTLs unless highly polymorphic markers are available for a large number of sib-pairs (Robertson, 1973). A number of modifications have been proposed to mitigate this limitation (e.g., Amos, Dawson, & Elston, 1990; Goldgar, 1990; Olson & Wijsman, 1993; Schork, Boehnke, Terwilliger, & Ott, 1993).

ANALYSIS OF DISCRETE TRAITS IN SIB-PAIRS

For traits that are multifactorial in nature, but measured on a discrete scale (e.g., disease states), powerful sib-pair approaches have been developed to test for linkage. In the case of discrete phenotypes, the test for linkage is a simple one-tailed *t*-test for a greater sharing of alleles ibd for affected over nonaffected individuals (Blackwelder & Elston, 1985). This method may also be used in situations in which genetic marker data on parents are unavailable. In this case, sib-pairs are scored according to the proportion of alleles that are shared "identity-by-state" (ibs); identity-by-state means that alleles are the same at a locus for the two siblings, irrespective of whether they are descended as copies from the same ancestral alleles (i.e., irrespective of whether they are ibd). With either ibd or ibs information, this approach shares the beneficial properties of the sib-pair methods for continuous traits; specifically, no assumptions concerning the mode of transmission and allowance for genetic heterogeneity are necessary. The general strategy is most useful when the sample comprises sib-pairs in which both members are affected (Blackwelder & Elston, 1985; Risch, 1990). Thus, this approach is quite advantageous for common diseases, but for less common conditions (e.g., schizophrenia), ascertainment of doubly affected pairs is difficult, thus limiting the practical situations in which this method may be applied.

Of course, the method also rests critically on accurate diagnosis of affection status.

SOME EXAMPLES OF QUANTITATIVE TRAIT LOCUS LINKAGE STUDIES

As mentioned previously, numerous pedigree and sib-pair studies have been undertaken to evaluate a wide range of medical and behavioral traits. As illustrative examples of such studies, we describe some of the results from ongoing studies of genetic loci for hypertension (see reviews by Williams et al., 1993, 1994). These studies of hypertension have followed a useful scientific model for genetic investigation, beginning with twin/family models to determine heritable traits relating to hypertension, then conducting appropriate segregation analyses to determine the mode of transmission, leading to linkage studies of pedigrees and sib-pairs, and, finally, identifying several functional mutations that contribute to hypertension. Several of the genetic loci identified represent striking examples of the greater complexity of multifactorial traits over monogenic systems, revealing gene–gene, gene–age, and gene–environment interactions.

Hypertension is a principal cause of cardiovascular morbidity and mortality, with a prevalence rate of 25–30% of Caucasian adults in the United States (Jeunemaitre, Lifton, Hunt, Williams, & Lalouel, 1992). The prevalence rate is considerably higher in African Americans (see Anderson, McNeilly, & Myers, 1992). It has been shown by a number of studies to aggregate in families and is regarded as etiologically heterogeneous and multifactorially determined (Ward, 1990). Several hypertension-related traits and their genetic etiologies are described in this volume (e.g., blood pressure in Chapters 5 and 6, body mass index in Chapters 7 and 8, and lipid subfractions in Chapter 12). Some additional characters and their heritabilities are presented in Table 1. As can be seen in the table, nearly all these traits are highly heritable, making them strong candidates for more fine-grained studies to map the loci responsible for their expression.

Hypertension is usually analyzed as a discrete trait in accordance with its disease classification. However, owing to its multifactorial inheritance, hypertension may not be optimal as a simple Mendelian marker in classic linkage studies (Ward, 1990; Lifton, Hunt, Williams, Pouyssegur, & Lalouel, 1991). The phenotypes in Table 1 allow further investigations of continuous variables that may be intermediate in the pathogenesis of hypertension (Lifton et al., 1991). Following the demonstration of heritable variation for these characters, segregation analyses have been conducted to determine the mode of transmission, which is necessary for pedigree studies of linkage (but unnecessary for sib-pair studies). The estimated heritabilities accounted for by the largest single gene for these traits are shown in Table 1 as "major locus heritabilities." All these are of sufficient size for detection using pedigree or sib-pair methods.

Williams et al. (1994) have summarized the genetic loci identified to date

TABLE 1. Some Hypertension-Related Traits and Their Heritabilities[a]

Trait	Heritability	
	Total	Major locus
High sodium–lithium countertransport	0.80	0.34
Low urinary kallikrein	0.78	0.51
High fasting insulin	0.44	0.33
Fat pattern index	0.52	0.42
Dense LDL subfractions	0.29	—
HDL cholesterol	0.45–0.74	—
Triglycerides	0.37–0.81	—
Body mass index	0.70	0.37

[a] Adapted from Williams et al. (1994). References for heritabilities are cited in the original publication.

that contribute to hypertension. These loci include angiotensinogen (AGT) on chromosome 1 (Jeunemaitre et al., 1992), lipoprotein-lipase on chromosome 8 (Emi et al., 1990), glucocorticoid remediable aldosteronism (GRA) on chromosome 8 (Lifton et al., 1991), low-density lipoprotein (LDL) receptor on chromosome 19 (Nishina et al., 1992), and insulin receptor on chromosome 19 (Ying et al., 1991). Other candidates suggesting linkage, but not confirmed as susceptibility genes for hypertension, include angiotensin-converting-enzyme on chromosome 17 (Zee et al., 1992; Jeunemaitre et al., 1992), renin on chromosome 1 (Zee et al., 1991), the human leukocyte antigen (HLA) locus on chromosome 6, and a sodium–hydrogen antiporter (Na^+-H^+) gene for sodium–lithium countertransport (SLC) on chromosome 1 (Lifton et al., 1991). The widely different functions and locations of these genes exemplify the multifactorial nature of hypertension.

With the possible exception of GRA, the susceptibility genes do not independently lead to hypertension. Rather, these genes interact among themselves and with the environment to express a number of phenotypes, one of which is hypertension. Williams et al. (1994) suggested that one set of genes, including AGT, NA^+-H^+, and an at present unknown gene for kallikrein production, interact with dietary intake of sodium, potassium, and other cations to induce essential hypertension. Another set of genes, including lipoprotein lipase (LPL), LDL, insulin receptor, and obesity genes, interact with physical (in)activity and diet. The gene for LPL deficiency exhibits a specific gene–age interaction. Under the age of 40, no clinical differences are apparent between individuals having and not having the deficiency variant of the LPL locus. Individuals older than 40, however, have significantly elevated levels of triglycerides, reduced levels of high-density lipoprotein (HDL) cholesterol, and significantly greater rates of hypertension (Wilson et al., 1990). Moreover, obese individuals appear more likely to express the lipid abnormalities in hypertension than do nonobese individuals. Interestingly, this is in complete accordance with the evidence for late-onset obesity-inducing genes discussed by Meyer in Chapter 8. Williams et al. (1994) suggested that hypertension onset may require two or three such genes from either or both categories plus environmental factors and that coronary heart disease may represent the culmination of the effects of the interactions among these susceptibility genes and environmental agents.

By using candidate loci in place of nontranscribing DNA segments as markers, the linkage studies of hypertension just described have the opportunity to move up to the next step of genetic analysis, that of identifying the functional mutations in the genes that lead to disease expression. The candidate loci approach is advantageous in this context because polymorphisms directly correspond to alternate forms of the gene. Thus, candidate loci allow direct investigations of associations between a specific gene form and the disease, thereby pinpointing the deleterious mutation and opening up possibilities for the development of biochemical screening tests. This also has the benefit of yielding to environmental intervention strategies that can compare the effects of specific environmental factors (e.g., controlled dietary intake of sodium, potassium, fat) on individuals genetically susceptible or resistant at a particular locus. This represents one exciting avenue of new research in hypertension epidemiology and etiology.

CONCLUSIONS

In this chapter, we have provided an overview of the issues and methods involved in mapping genes for complex traits. Clearly, the tools required and obstacles encountered in such analyses differ from those in classic twin, family, and adoption studies of behavior. However, as we have attempted to demonstrate by discussion of sib-pair designs and applications to hypertension, these different approaches represent two successive, complementary steps toward precise identification and characterization of the causes of variability in behavioral and medical phenotypes. Both steps are essential for a thorough understanding of the etiology of complex traits.

The sib-pair design for QTL linkage analysis corresponds well to the classic twin study, both in design and in substance. Given a study of behavior in twins, for example, all that is needed beyond the initial phenotypes are DNA marker data drawn from blood samples. Since DZ twins are, genetically, full siblings, all the sib-pair methods apply equally well to either type of sample. If candidate loci for the trait of interest are available, as in the case of hypertension or obesity, for example, then the task of marker acquisition may be undertaken quite simply and efficiently. If there are no immediate candidates, however, one may still conduct a QTL study by collecting markers that span the genome and then conduct linkage tests for all markers. Due to advances in molecular genetic technology, the costs of a (rough) genome search are no longer prohibitive. In fact, at least one large-scale study of this kind is at present under way (Plomin et al., 1994). In this manner, classic studies of genetic and environmental influences in twins or other family members allow a very simple extension to identifying the actual elements that comprise the inferred genetic effects.

Finally, it is also important to recognize that in research on QTL linkage, as in other research using twin or family data, the designs of the studies and the results from their analysis offer unique opportunities for greater understanding of environmental factors that contribute to individual differences. In the twin or family designs, we can account for genetic influences and thereby evaluate important features of the environment unencumbered by genetic confounds. Similarly, by locating and characterizing variations in functional genes, we can examine the environmental agents that trigger their expression and ultimately lead to observable phenotypes. Thus, by mapping the genes for complex traits, we can begin to piece together the puzzle of the interplay of nature and nurture in determining our behavior.

REFERENCES

Amos, C. I., Dawson, D. V., & Elston, R. C. (1990). The probabilistic determination of identity-by-descent sharing for pairs of relatives from pedigrees. *American Journal of Human Genetics, 47*, 842–853.

Anderson, N. B., McNeilly, M., & Myers, H. (1992). Toward understanding race difference in autonomic reactivity: A proposed contextual model. In J. R. Turner, A. Sherwood, & K. C. Light (Eds.), *Individual differences in cardiovascular response to stress* (pp. 125–145). New York: Plenum Press.

Blackwelder, W. C., & Elston, R. C. (1985). A comparison of sib-pair linkage tests for disease susceptibility loci. *Genetic Epidemiology, 2*, 85–97.

Bodmer, W. F., Bailey, C. J., Bodmer, J., Bussey, H. J. R., Ellis, A., Gorman, P., Lucibello, F. C., Murday, V. A., Rider, S. H., & Scambler, P. (1987). Localization of the gene for familial adenomatous polyposis on chromosome 5. *Nature, 328*, 614–616.

Cannon-Albright, L. A., Goldgar, D. E., Meyer, L. J., Lewis, C. M., Anderson, D. E., Fountain, J. W., Hegi, M. E., Wiseman, R. W., Petty, E. M., & Bale, A. (1992). Assignment of a locus for familial melanoma, MLM, to chromosome 9p130p22. *Science, 258*, 1148–1152.

Cardon, L. R., Smith, S. D., Fulker, D. W., Kimberling, W. J., Pennington, B. F., & DeFries, J. C. (1995). Quantitative trait locus on chromosome 6 predisposing to reading disability. *Science, 266*, 276–279.

Cardon, L. R., & Fulker, D. W. (1994). The power of interval mapping of quantitative trait loci using selected sib pairs. *American Journal of Human Genetics, 55*, 825–833.

Carey, G., & Williamson, J. (1991). Linkage analysis of quantitative traits: Increased power by using selected samples. *American Journal of Human Genetics, 49*, 786–796.

DeFries, J. C., & Fulker, D. W. (1985). Multiple regression analysis of twin data. *Behavior Genetics, 15*, 467–473.

DeFries, J. C., & Fulker, D. W. (1988). Multiple regression analysis of twin data: Etiology of deviant scores versus individual differences. *Acta Geneticae Medicae et Gemellalogiae, 37*, 205–216.

DeFries, J. C., Fulker, D. W., & LaBuda (1987). Evidence for a genetic aetiology in reading disability of twins. *Nature, 329*, 537–539.

Emi, M., Hata, A., Robertson, M., Iverius, P. H., Hegele, R., Lalouel, J. M. (1990). Lipoprotein lipase deficiency resulting from a nonsense mutation in exon 3 of the lipoprotein lipase gene. *American Journal of Human Genetics, 47,* 107–111.

Falconer, D. S. (1990). *Introduction to quantitative genetics,* 3rd ed. New York: Longman Group.

Fulker, D. W., & Cardon, L. R. (1994). A sib pair approach to interval mapping of quantitative trait loci. *American Journal of Human Genetics, 54,* 1092–1103.

Fulker, D. W., Cardon, L. R., DeFries, J. C., Kimberling, W. J., Pennington, B. F., & Smith, S. D. (1991). Multiple regression analysis of sib-pair data on reading to detect quantitative trait loci. *Reading and Writing: An Interdisciplinary Journal, 3,* 299–313.

Fulker, D. W., Cherny, S. S., & Cardon, L. R. (1995). Multipoint interval mapping of quantitative trait loci using sib pairs (submitted).

Gelderman, H. (1975). Investigations on inheritance of quantitative characters in animals by gene markers. I. Methods. *Theoretical Applied Genetics, 46,* 319–330.

Gillis Light, J., & DeFries, J. C. (1994). Comorbidity of reading and mathematics disabilities: Genetic and environmental etiologies. *Journal of Learning Disabilities* (in press).

Goldgar, D. E. (1990). Multipoint analysis of human quantitative genetic variation. *American Journal of Human Genetics, 47,* 957–967.

Haseman, J. K., & Elston, R. C. (1972). The investigation of linkage between a quantitative trait and a marker locus. *Behavior Genetics, 2,* 3–19.

Hill, W. G. (1974). Estimation of linkage disequilibrium in randomly mating populations. *Heredity, 33,* 229–239.

Innis, M. A., Gelfand, D. H., Sninskey, J. J., & White, T. J. (1990). *PCR protocols: A guide to methods and applications.* New York: Academic Press.

Jeunemaitre, X., Lifton, R. P., Hunt, S. C., Williams, R. R., & Lalouel, J.-M. (1992). Absence of linkage between the angiotensin-converting enzyme and human essential hypertension. *Nature and Genetics, 1,* 72–75.

Lifton, R. P., Hunt, S. C., Williams, R. R., Pouyssegur, J., & Lalouel, J.-M. (1991). Exclusion of the Na^+-H^+ antiporter as a candidate gene in human essential hypertension. *Hypertension, 17,* 8–14.

Mather, K., & Jinks, J. L. (1971). *Biometrical genetics.* London: Chapman & Hall.

Neale, M. C., & Cardon, L. R. (1992). *Methodology for genetic studies of twins and families.* Boston: Kluwer Academic Publishers.

NIH/CEPH Collaborative Mapping Group. (1992). A comprehensive genetic linkage map of the human genome. *Science, 258,* 67–86.

Nishina, P. M., Johnson, J. P., Naggert, J. K., Krauss, R. M. (1992). Linkage of atherogenic lipoprotein phenotype tot he low density lipoprotein receptor locus on the short arm of chromosome 19. *Proceedings of the National Academy of Sciences, USA, 89,* 708–712.

Olson, J. M., & Wijsman, E. M. (1993). Linkage between quantitative trait and marker loci: Methods using all relative pairs. *Genetic Epidemiology, 10,* 87–102.

Ott, J. (1991). *Analysis of human genetic linkage.* Baltimore: Johns Hopkins University Press.

Penrose, L. S. (1938). Genetic linkage in graded human characters. *Annals Eugenics, 6,* 133–138.

Plomin, R., McClearn, G. E., Smith, D. L., Vignetti, S., Chorney, M. J., Chorney, K., et al. (1994). DNA markers associated with high versus low IQ: The IQ quantitative trait loci (QTL) project. *Behavior Genetics, 24,* 107–118.

Risch, N. (1990). Linkage strategies for genetically complex traits. II. The power of affected relative pairs. *American Journal of Human Genetics, 46,* 229–241.

Robertson, A. (1973). Linkage between marker loci and those affecting a quantitative trait. *Behavior Genetics, 3,* 389–391.

Schork, N. J., Amos, C. I., Fulker, D. W., & Cardon, L. R. (1995). Interval mappiing of quantitative trait loci using variance components (submitted).

Schork, N. J., Boehnke, M., Terwilliger, J. D., & Ott, J. (1993). Two-trait-locus linkage analysis: A powerful strategy for mapping complex genetic traits. *American Journal of Human Genetics, 53,* 1127–1136.

Schork, N. J., & Weder, A. B. (1995). Statistical analysis of genetic loci influencing quantitative traits that exhibit postnatal developmental patterns (submitted).

Ward, R. (1990). Familial aggregation and genetic epidemiology of blood pressure. In J. H. Laragh & B. M. Brenner (Eds.), *Hypertension: Pathophysiology, diagnosis and management* (pp. 81–100). New York: Raven Press.

Williams, R. R., Hunt, S. C., Hopkins, P. N., Hasstedt, S. J., Wu, L. L., & Lalouel, J.-M. (1994). Tabulations and expectations regarding the genetics of human hypertension. *Kidney International, 44,* S57–S64.

Williams, R. R., Hunt, S. C., Hopkins, P. N., Wu, L. L., Hasstedt, S. J., Berry, T. D., Barlow, G. K., Stults, B. M., Schumacher, M. C., Ludwig, E. H., Elbein, S. C., Wilson, D. E., Lifton, R. P., & Lalouel, J.-M. (1993). Genetic basis of familial dyslipidemia and hypertension: 15-Year results from Utah. *Hypertension, 6,* 319S–327S.

Wilson, D. E., Emi, M., Iverius, P. H., Hata, A., Wu, L. L., Hillas, E., Williams, R. R., & Lalouel, J.-M. (1990). Phenotypic expression of heterozygous lipoprotein lipase deficiency in the extended pedigree of a proband homozygous for a missense mutation. *Journal of Clinical Investigation, 86,* 735–750.

Ying, L. J., Zee, R. Y., Griffiths, L. R., & Morris, B. J. (1991). Association of a RFLP for the insulin receptor gene, but not insulin, with essential hypertension. *Biochemistry and Biophysics Research Communication, 181,* 486–492.

Zee, R. Y., Ying, L. H., Morris, B. J., & Griffiths, L. R. (1991). Association and linkage analyses of restriction fragment length polymorphisms for the human renin and antithrombin III genes in essential hypertension. *Journal of Hypertension, 9,* 825–830.

Zee, R. J., Lou, Y. K., Griffiths, L. R., & Morris, B. J. (1992). Association of a polymorphism of the angiotensin I-converting enzyme gene with essential hypertension. *Biochemistry and Biophysics Research Communication, 184,* 9–15.

PART SEVEN

EPILOGUE

CHAPTER FOURTEEN

Implications of a Behavior Genetic Perspective in Behavioral Medicine

JOHN K. HEWITT AND J. RICK TURNER

The contributors to this volume have provided detailed accounts of current behavior genetic research on alcoholism and smoking as examples of addictive behaviors, on cardiovascular parameters important to our understanding of stress responses, and on eating disorders and anthropometric indices related to obesity. While these topics are by no means exhaustive, they do provide central and substantive examples of behavior genetic research in behavioral medicine. Beyond this focus on specific phenotypes, our contributors have also considered the more general issues of gender differences in genetic and environmental influences and the relationship between behavior genetics and health-related research in minority populations.

Last, we were taken in two further directions. In one, the nature of environmental variation was questioned. Specifically, it was indicated that some aspects of what we normally consider to be environmental variation might themselves be heritable. In the other direction, new methods for locating the individual genes that contribute to the genetic influences on complex traits were described and their relevance to behavioral medicine established.

Thus, we have covered a lot of ground that has offered us both detailed presentations of empirical research and challenging new ideas. To conclude, we would like to emphasize three general themes that characterize all the contribu-

JOHN K. HEWITT • Institute for Behavioral Genetics, University of Colorado at Boulder, Boulder, CO 80309. J. RICK TURNER • Departments of Pediatrics and Preventive Medicine, University of Tennessee, Memphis, TN 38105; *present address:* Department of Pediatrics and Georgia Prevention Institute, Medical College of Georgia, Augusta, Georgia 30912.

Behavior Genetic Approaches in Behavioral Medicine, edited by J. Rick Turner, Lon R. Cardon, and John K. Hewitt. Plenum Press, New York, 1995.

tions to this volume and that highlight the implications of a behavior genetic perspective in behavioral medicine.

The first theme is that in order to understand the role of genetic influences, we must pay close attention to the phenotypes we are studying and also to the environmental, cultural, or other contexts in which genetic influences are expressed. It is perhaps this aspect of behavior genetics that sets it aside most clearly from other kinds of genetic research.

The characteristics that we study are dynamic. They change during individual biological development (e.g., blood pressure, body mass index, and adiposity), in response to environmental stresses (as exemplified by cardiovascular reactivity), and through individual life courses or social changes (as with addictive behaviors). Equally, these characteristics are complex, in the sense that they involve several components, such as the initial exposure, experimentation, regular use, and attempts to quit that characterize smoking behavior, and the changes in cardiac output and peripheral vascular resistance that determine blood pressure responses to stress.

Perhaps because of these attributes of dynamic change and complexity that are typical of characteristics studied by behavioral geneticists, the contributions to this volume seem a far cry from the misrepresentations of behavior genetic research that have sometimes surfaced. As the contributions to this volume amply demonstrate, it is just not true that the causes of individual differences are *either* environmental *or* genetic. In almost every case, *both* genetic *and* environmental variation are found to be important, and the questions of interest then concern how these influences operate and how they are changed in different circumstances.

A particularly compelling illustration of this fact came from the study of addictive behaviors (Chapters 2, 3, and 4). For smoking, it is apparent that the process of initiation is under different genetic influences and environmental influences (both individual and social) than is smoking persistence. Among other things, smoking initiation in most populations is determined to a greater extent by shared family, peer, and sibling environmental influences than are differences among individuals in their smoking persistence. Furthermore, the relative importance of genetic and environmental influences—on smoking initiation, for example—may differ in men and women and across cultures. These differential patterns of genetic and environmental influences on different components of the behavior suggest particular hypotheses about the process of transition from nonsmoker to regular smoker in a given cultural context (see Chapter 4 by Rowe and Linver). The different mediation of genetic influences on smoking initiation and on smoking persistence also provides insights into what kinds of factors make individuals most vulnerable to developing a habit that will be significantly health-threatening. Furthermore, as Heath and Madden (Chapter 3) emphasize, genetic studies have raised the critical question of the extent to which any genetic vulnerability to develop addictions is substance-specific or, alternatively, is general to alcohol, nicotine, and other substances. Multivariate genetic studies will answer this question.

A second common theme of the research presented and reviewed in this volume is that genetic factors *do* influence the characteristics under study. This is true for addictive behaviors, stress responses, eating disorders, adiposity and the body mass index, and even for variation in events and experiences that have traditionally been treated as purely environmental (see Chapter 12 by Plomin). As we have seen, the degree of genetic influence and the mode of that influence are different for each phenotype, perhaps for each component of that phenotype, and probably vary with the circumstances in which the genes are expressed.

In some areas, such as individual differences in the body mass index, genetics may turn out to be the single most important factor. As we noted in Chapter 1, this finding has the implication that we might want to turn, in this case, to the search for individual genes that, through their influence on metabolism, appetite control, energy expenditure, and so on, exert control over our body composition. Indeed, as outlined in Chapter 13, the necessary methodology for locating these quantitative trait genes is now becoming available.

The third common theme of these contributions is the one that we hope will become the"take home message" from this volume. Behavior genetic methods are readily accessible, and they are applicable to the phenotypes of interest in behavioral medicine. The insights that a behavior genetic perspective provides will complement, not replace, those from other perspectives. As we saw in Chapters 10 and 11, behavior genetics can also enhance, and is enhanced by, the study of gender and minority group issues, and it does not bring with it any preconceptions about the causes of individual differences.

We hope that readers of this volume, particularly those who might previously have been unfamiliar with behavior genetic research in behavioral medicine, will share the excitement we now feel about the prospects for this collaborative exercise in scientific discovery.

Author Index

Subject Index

MIX
Papier aus verantwortungsvollen Quellen
Paper from responsible sources
FSC® C105338

If you have any concerns about our products,
you can contact us on
ProductSafety@springernature.com

In case Publisher is established outside the EU,
the EU authorized representative is:
Springer Nature Customer Service Center GmbH
Europaplatz 3, 69115 Heidelberg, Germany

Printed by Libri Plureos GmbH
in Hamburg, Germany